Precalculus
Pathways to Calculus
A Problem Solving Approach
Student Workbook

Ninth Edition

Marilyn P. Carlson
Arizona State University

Alan E. O'Bryan
Arizona State University

Michael Oehrtman
Oklahoma State University

Kevin Moore
University of Georgia

Cover images © Shutterstock, Inc.

Precalculus: Pathways to Calculus – Student Workbook, Ninth Edition

www.kendallhunt.com
Send all inquiries to:
4050 Westmark Drive
Dubuque, IA 52004-1840

Copyright © 2022 by Great River Learning

Text + website ISBN 978-1-68478-086-0
Text ISBN 978-1-68478-017-4

Published in the United States of America

Introduction

Overview of Workbook Content

This workbook contains investigations and homework for the text's eleven modules. The investigations introduce the modules' central ideas and include questions to help you build your own understanding of these ideas. Each module in this workbook has three to ten investigations and one homework set. The sections of the homework follow the naming and ordering of the investigations and text. The text contains online videos that explain key ideas and explain examples like ones in your homework—we urge you to use these vides to help you complete your homework and strengthen your understandings. We have intentionally included problems in each section of the homework that are much like problems on the investigations and in the videos. This is so you can develop the thinking and understandings that will be needed to be successful in calculus.

The first module focuses on understanding and practicing methods of simplifying expressions, evaluating functions and solving equations. This module has three investigations that provide review of these important skills. You can also find many practice problems in Module 1 of the online text. All you need to do is just follow the links.

Modules 2 – 11 focus on building your understanding of standard precalculus ideas that are foundational for calculus. By engaging with the questions in the investigations and homework your reasoning and problem solving abilities will get better and better. Over time you will become powerful mathematical thinker who has confidence in your ability to figure out novel problems on your own.

To: The Precalculus Student

Welcome!

You are about to begin a new mathematical journey that we hope will lead to your choosing to continue studying mathematics. Even if you don't currently view yourself as a math person, it is very likely that these materials and this course will change your perspective. The materials in this workbook are designed with student learning and success in mind and are based on decades of research on student learning. In addition to becoming more confident in your mathematical abilities, the reasoning patterns, problem solving abilities and content knowledge your acquire will make more advanced courses in mathematics, the sciences, engineering, nursing, and business more accessible. The investigations and text will help you see a purpose for learning and understanding the ideas of precalculus, while also helping you acquire critical knowledge and ways of thinking that you will need for learning calculus. To assure your success, we urge you to advantage of the many resources we have provided to support your learning. We also ask that you make a strong effort to make sense of the questions and ideas that you encounter. This will assure that your mathematical journey through this course is rewarding and transformational.

Wishing you much success!

Dr. Marilyn P. Carlson, Dr. Alan E. O'Bryan, Dr. Michael Oehrtman, Dr. Kevin Moore,
and the Pathways Web Developer and Manager, Tim Persson

Table of Contents_____

This module develops foundational ideas of angle and angle-measure by investigating approaches for measuring the openness of two rays. We introduce methods for modeling the behavior of periodic motion in the context of co-varying an angle measure with a linear measurement that maps out a periodic motion, laying the groundwork for introducing the trigonometric functions of sine and cosine. The module concludes by exploring the meaning of period, amplitude, and translations of both the sine and cosine functions.

We explain the relationship between right triangle and unit circle trigonometry by initially exploring the right triangle relationships defined by sine, cosine, and tangent functions in a unit circle context. The triangle relationships defined by the sine, cosine, and tangent functions are used to determine the values of unknown quantities in various applied problems. We conclude the module by deriving various trigonometric identities that relate the trigonometric functions to one another.

We introduce polar coordinates with examples that illustrate the usefulness of polar coordinates as an alternate coordinate system. Students practice graphing polar functions and using polar functions to track the co-variation of distances and directions to plot and interpret the graphs of these functions. We next spend time exploring how constant rate of change is represented in polar coordinates and how to identify the graphs of functions with a constant rate of change. Similarly, we explore the meaning of the average rate of change for functions defined in polar coordinates. After working with average rates of change, we determine methods for converting between rectangular and polar coordinates and converting functions defined in one system to a function in the other system that produces the same graph.

This module begins by exploring what a vector is and how to represent a vector in both polar form and component (or rectangular) form. We next investigate vector addition as a way of determining the net effect of multiple vectors interacting with one another. We conclude the module by learning how to scale vectors (the process of changing the magnitude of a vector by some factor) with both positive and negative scalars.

We begin this module by introducing terminology and then learn how to graph sequences by generating ordered pairs consisting of the position and value of the terms. We introduce notation for describing sequences and how to define patterns using both recursive and explicit formulas. We apply these ideas by developing the meaning of arithmetic and geometric sequences and develop formulas to represent them. We then define what we mean by a series and use sequences of partial sums to keep a running total of the sum of the terms of the sequence. The module concludes by establishing sigma notation as a way of representing a series and by developing methods for finding the sums of finite arithmetic series, finite geometric series, and infinite geometric series.

Simplify or Expand to Produce Equivalent Expressions

Order of Operations

1. Evaluate the following:

   a. $-2+5-12$

   b. $6-(-4)+1$

   c. $\dfrac{-5+(9-5(4))+3}{5(-2)-1}$

2. Evaluate the following:

   a. $\left|-5-(-2)\right|$

   b. $\left|\dfrac{-3(5-1)}{-4}\right|$

3. Suppose that $a=2$, $b=-3$, $c=0$, and $d=-5$. Find the value of the following expressions.

   a. $b(2a-3d)$

   b. d^2-25+c

   c. $6-\dfrac{b}{ad}$

   d. $-a\sqrt{b^2+c^2}$

4. a. Simplify the following.

   i. $\dfrac{(2)(2)(2)(2)(2)}{(2)(2)(2)}$

   ii. $\dfrac{(7)(7)(7)}{(7)(7)(7)}$

   iii. $\dfrac{(5)(5)}{(5)(5)(5)(5)(5)}$

   iv. $\dfrac{(x-3)(x-3)}{(x-3)}$

   v. $\dfrac{(x+2)^2(x-1)}{(x+2)^5(x-1)(x+1)}$

   vi. $\dfrac{x^3y^2z^2}{x^5yz^4}$

   b. Explain why $\dfrac{x^m}{x^n}=x^{m-n}$.

   c. Expand and multiply to justify why $\left(x^3\right)^5=x^{15}$, then explain why $\left(x^m\right)^n-x^{(m)(n)}$.

   d. Simplify: $\dfrac{96x^6y^7z^9}{128x^6(y^3z)^4}$

   e. Explain why $x^0=1$.

5. Simplify the following. Verify that your simplified expression is equivalent to the original expression by substituting several different values for x. If the expression does not simplify, write DNS.

   a. $2(x^2+1)$

   b. $-2(x-5)$

   c. $-2x(5x^2+4)$

   d. $\dfrac{-15x+7}{3}$

   e. $\dfrac{6x^4-5}{-2x}$

   f. $\dfrac{-6x^5+7x}{-3x}$

6. Are the following statements valid? Explain and verify by substituting numbers in for the variables and evaluating.

   a. $4y=4x+5$ implies that $y=x+5$

   b. $\dfrac{ab+c}{a}=b+c$

   c. $-5a(-7a+5)=35a^2-25a$

   d. $\dfrac{2y-6x}{3x}=2y-2$

7. Simplify: $\dfrac{1}{5}\left(30x^4+12x^2-3x-5\right)-\left(2x^4-6x^2+5x\right)$

8. Multiply each of the following to obtain equivalent expressions.

   a. $(x-3)(x+9)$

   b. $(2x-7)(-3x+2)$

   c. $(x+5)^2$

   d. $(3m-2n)^2$

   e. $(3x-4y)(5z+8x^2)$

9. Draw a square and label the side length, $s=4$ kilometers. Illustrate why we use the formula $A=s^2$ to compute the area of a square. What is the unit of the answer? Explain.

10. Illustrate on this diagram why the area of a square with side length $a+b$ units is equal to $a^2 + 2ab + b^2$ square units.

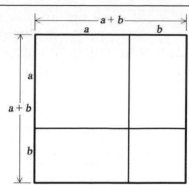

11. Is it true that $(a+b)^2 = a^2 + b^2$? Explain.

12. Factor each of the following to obtain equivalent expressions.

    a. $x^2 - 2xy + y^2$ b. $3a^2 + 6ab + 3b^2$

    c. $2x^3 + 5x^2 - 12x$ d. $16x^2 - 36$

13. Simplify each of the following expressions. If the expression does not simplify, write DNS.

    a. $\dfrac{x^2 + 5x - 14}{x-2}$ b. $\dfrac{x+1}{3x^2 - x - 4}$

    c. $\dfrac{x^3}{x^3 - 2x^2 + 5x}$ d. $\dfrac{x^2 - 7x + 2}{x}$

14. Simplify the following radicals to obtain equivalent expressions. If the expression does not simplify, write DNS.

    a. $\sqrt{75x^9 y^2 z}$ b. $\sqrt{\dfrac{a^{11}}{b^4 c^7}}$ c. $\sqrt[3]{24a^5 b^2 c^3}$

    d. $\sqrt{a^2 - b^2}$ e. $\sqrt{a^2 + 9}$ f. $\sqrt{a^2 + 2ab + b^2}$

15. A student has simplified the following expressions **incorrectly**. Identify the student's error and then substitute numbers in for the variables to verify that the expressions are not equivalent. **Correct the error by writing an expression that is equivalent to the one on the left side of the equal sign.**

a.

$$(2a - 3b)^2 = 4a^2 - 9b^2$$

b.

$$3x(x-2) - 4(5-x) = 3x^2 - 6x - 20 - x$$
$$= 3x^2 - 7x - 20$$

c.

$$3(2x-5)^2 = (6x-15)^2$$
$$= 36x^2 - 180x + 225$$

d.

$$\frac{25x^3 - x}{5x} = 5x^2 - x$$

e.

$$\sqrt{x^2 + 16} = x + 4$$

f.

$$\sqrt{x^2 - 25} = x - 5$$

g.

$$\sqrt{-18} = -3\sqrt{2}$$

1. a. Indicate on the figure and describe the attribute of the rectangle that represents the following quantities.
       l = length of the side of the rectangle measured in inches
       w = width of the side of the rectangle measured in inches
       p = perimeter of the rectangle measured in inches
       A = area of the rectangle measured in square inches

    b. Write the formula that determines the perimeter of a rectangle, given its width and length.

    c. Determine the width w of a rectangle when $p = 21$ inches and $l = 7$ inches. Show your work.

    d. Write the formula that determines the area of a rectangle, given its width and length.

    e. Determine the area A of a rectangle when $p = 26$ inches and $l = 5$ inches. Show your work.

2. a. Indicate on the figure and describe the attribute of the circle that represents the following quantities.
       r = radius of the circle measured in feet
       d = diameter of the circle measured in feet
       C = circumference of the circle measured in feet
       A = area of the circle measured in square feet

    b. Write the formula to express a circle's diameter in terms of its radius. Use the formula to determine the diameter of a circle that has a radius of 4.721 feet.

    c. Write the formula to express a circle's circumference in terms of its diameter. Use the formula to determine the circumference of a circle that has a diameter of 6.48 feet.

    d. Write the formula to express a circle's area in terms of its diameter. Use the formula to determine the area of a circle that has a diameter of 4.09 feet.

    e. Write the formula to express a circle's circumference in terms of its radius. Use the formula to determine the circumference of a circle that has a radius of 3.5 feet.

f. Write the formula to express a circle's area in terms of its diameter. Use the formula to determine the area of a circle that has a diameter of 3.5 feet.

g. Define a formula to express the circumference of a circle in terms of its area. Use the formula to determine the circumference of a circle that has an area of 42.7 square feet.

3. a. Indicate on the figure and describe the attribute of the cube that represents the following quantities.
   s = length of the sides of the cube measured in centimeters
   SA = surface area of the cube measured in square centimeters
   V = volume of the cube measured in cubic centimeters

   b. Write the formula to express the volume of a cube V in terms of its side length s. Determine the volume of a cube that has a side length of $4\frac{5}{8}$ centimeters.

   c. Write the formula to express the surface area of a cube SA in terms of its side length s. Determine the surface area of a cube when the cube's side length is $2\frac{2}{3}$ centimeters.

   d. Define a formula to express the volume of a cube V in terms of its surface area SA. Determine the volume of a cube that has a surface area of 62 square centimeters.

4. a. How much greater is the area of a square with a side length of 8 inches than the area of a circle with a radius of 3 inches?

   b. Make a drawing to illustrate a circle with a radius of 3 inches inside a square with a side length of 8 inches and shade the area that represents the difference between these two areas.

   c. When the area of a square with side length l is larger than the area of a circle with radius r, define a formula to express the difference between the two areas. What is the side length of the square, if the difference in area is 2 square inches and the radius of the circle is 3 inches?

1. A candle that is originally 10 inches long has burned x inches.
   a. Make a drawing of the 10-inch candle and illustrate the quantities 10 inches, $10 - x$ inches, and x inches on your drawing. What does $10 - x$ represent?

   b. Define a formula to express the length of the candle y in terms of the number of inches that have burned from the candle x.

   c. Determine the value of x when the length of the candle y is 3.4 inches. What does the solution to this equation represent?

   d. Construct a graph that represents the length of the candle y in terms of the number of inches that have burned from the candle x. Label two points on your graph and describe what each point represents.

   e. As the length of the candle decreases, how does the number of inches that have burned from the candle change?

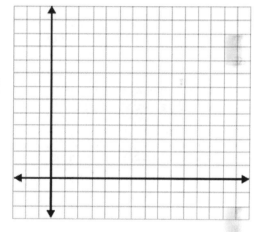

   f. Set up an equation to algebraically determine the number of inches that have burned from the candle when the length of the candle is 1.4 inches. Illustrate your solution on your graph.

   g. Describe what it means to "solve for x" in the context of this problem.

2. Write an equation for each of these situations and then *solve the equation* for the unknown value. Define a variable to represent the value of the unknown number before writing each equation.
   a. 17.5 is equal to 2 times some number. What is the number?

   b. The sum of 3 times some number and 12 is 42. What is the number?

   c. $\frac{1}{4}$ of some number is 4.3. What is the number?

d.   The difference between some number and 14.3 is 2.1. What is the number?

e.   Some number is 4 times as large as 9.8. What is the number?

f.   Some number is equal to $\frac{1}{3}$ of the sum of 88.2, 93.5 and 64. What is the number?

g.   45 is some multiple of 15. Set up an equation and determine the value of the multiple.

h. $1,200 is 1.5 times as large as some number. What is the number?

i.   200% of some number is 38.2. What is the number?

j. $\frac{3}{4}$ of some number increased by 12 is equal to 5 times the number. What is the number?

3.   The distance between some number(s) and 0 on the number line is 4.
   a.   What are the numbers?

   b.   Illustrate on the number line all numbers that are 4 units away from 0.

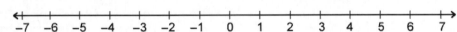

   c.   Solve the equation $|x| = 4$ for x (what values of x make this equation true)?

   d.   Describe how the solutions found in part (c) are represented on the number line.

4.   Solve the following equations for the unknown value(s) (i.e., solve for x).
   a.   Given that $|x| = 5$, determine the value(s) of x that makes this equation true.

   b.   Given that $2x - 19 = 5$, what value(s) of x makes this equation true.

   c.   Given that $\sqrt{x} = 4$, what value(s) of x makes this equation true.

d. Given that $x^2 = 36$, what value(s) of x makes this equation true.

e. Given that $\sqrt[3]{x} = -2$, what value(s) of x makes this equation true.

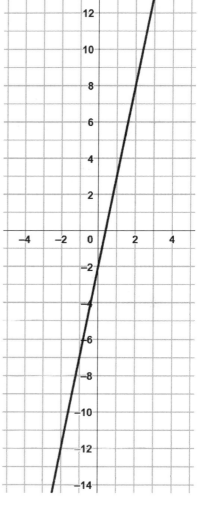

5. a. Given a graph that represents how x and y are related, determine the value of x when $y = 8$. Describe your approach and say why it works.

b. Given that $y = 5x - 2$ is the equation of the line graphed, solve the equation $8 = 5x - 2$ for x algebraically. Illustrate how to use the graph to solve this equation.

c. Use the graph to solve the equation $3 = 5x - 2$. Illustrate how you used the graph to solve this equation.

d. Use the graph to solve the equation $0 = 5x - 2$. Illustrate how you used the graph to solve this equation.

6. Solve the following equations for x:

   a. $\dfrac{3}{4}x + \dfrac{5}{6} = 5x - \dfrac{125}{3}$

   b. $\sqrt{x+1} - 3x = 1$

   c. $|5x - 6| + 3 = 10$

   d. $4 = \dfrac{x+3}{x-1}$

7. A student has **incorrectly** solved the following equations for x when $y = 0$. Describe the mistake(s) made by the student. Then provide a **correct** solution for each of the equations.

   a.

   $$y = -5 + \sqrt{x^2 + 16} \quad ; \quad y = 0$$

   Solution: $0 = -5 + \sqrt{x^2 + 16}$
   $0 = -5 + x + 4$
   $0 = -1 + x$
   $\boxed{1 = x}$

   b.

   $$y = x^3 - 3x^2 - 4x \quad ; \quad y = 0$$

   Solution: $0 = x^3 - 3x^2 - 4x$
   $\dfrac{0}{x} = \dfrac{x^3}{x} - \dfrac{3x^2}{x} - \dfrac{4x}{x}$
   $0 = x^2 - 3x - 4$
   $0 = (x + 4)(x - 1)$
   Thus, $x + 4 = 0 \qquad x - 1 = 0$
   $\boxed{x = -4} \qquad \boxed{x = 1}$

This investigation contains review and practice with important skills and procedures you may need in this module and future modules. Your instructor may assign this investigation as an introduction to the module or may ask you to complete select exercises "just in time" to help you when needed. Alternatively, you can complete these exercises on your own to help review important skills.

Order of Operations
Use this section prior to the module or with/after Investigation 1.

The order of operations is an agreed-upon convention that ensures expressions are evaluated the same way by everyone. The commonly accepted order is as follows.

   i. *Simplify any expressions contained within parentheses or brackets. If there are nested sets of parentheses, then work from the inside to the outside.*
   ii. *Evaluate any parts of the expression containing exponents.*
   iii. *Negate any terms with a negation symbol in front of them.*
   iv. *Perform all multiplication and division, working from left to right.*
   v. *Perform all addition and subtraction, working from left to right.*

In Exercises #1-3, use the order of operations to simplify each expression.

1. $3 + 6(4) \div 12 - 1$
2. $4(3+2)^2 - 10$
3. $-(16 - 2^3) + 4$

When an expression is written as a fraction, we think of the entire numerator and denominator as having parentheses around them. Thus, we simplify the entire numerator as much as possible and the entire denominator as much as possible before reducing or evaluating the quotient.

In Exercises #4-6, use the order of operations to simplify each expression.

4. $\dfrac{13 - 3^2}{2^3}$
5. $\dfrac{4 - 6 \div 2}{10 + 3 \cdot (-3)}$
6. $\dfrac{-5 + (9 - 5(4)) + 3}{5(-2) - 1}$

In Exercises #7-9, place parentheses to guide the order of operations so the expressions evaluate to the indicated value. For example, if we want the expression $3 + 6 \cdot 4 - 2$ to evaluate to 18 we need parentheses as follows: $(3+6) \cdot (4-2)$. Any other way of placing the parentheses, or evaluating it as its written, will produce a different result.

7. $3 + 6 \cdot 4 - 2$ so that its value is 34
8. $5 - 4 \cdot 2^2 + 18$ so that its value is 22

9. $5 - 4 \cdot 2^2 + 18$ so that its value is -41

The Distributive Property
Use this section prior to the module or with/after Investigation 2.

The distributive property is stated as follows. For real numbers a, b, and c,

$a(b + c) = ab + ac$ [can also be written $ab + ac = a(b + c)$]

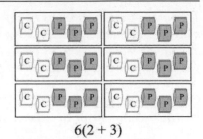

$6(2 + 3)$

The distributive property is all about grouping. Suppose six friends each bring two bags of tortilla chips and three bags of pretzels to stock the snack bar at a dorm movie night. How many bags of snacks did they bring?

We can think of this in two ways. The first method is to count the number of bags of snacks each person brings [(2 + 3), or 5] and multiply this by the number of people [6]. So the total number of bags of snacks [30] can be represented as shown to the right (top).

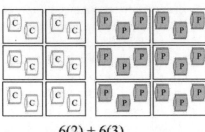

$6(2) + 6(3)$

The second method is to count the number of bags of tortilla chips [6(2), or 12] and the number of bags of pretzels [6(3), or 18] and add them together [12 + 18]. So the total number of bags of snacks [30] can be represented as shown to the right (bottom).

Therefore, we can understand why the expressions $6(2 + 3)$ and $6(2) + 6(3)$ are equivalent (beyond the fact that they evaluate to the same number).

In Exercises #10-12, use the distributive property to rewrite the expression in expanded form. For example, we can rewrite $3(6 + x)$ as $3(6) + 3x$ or $18 + 3x$.

10. $2(5 + r)$ 11. $7(3x + 2)$ 12. $x(7 + 4y)$

When subtraction is involved, you might choose to change the expression to involve addition. This may help you avoid sign errors (including forgetting about the negative). For example, you can rewrite $2(x - 5)$ as $2(x + (-5))$. Then applying the distributive property yields $2x + 2(-5)$, or $2x + (-10)$, or $2x - 10$.

In Exercises #13-15, use the distributive property to rewrite the expression in expanded form.

13. $6(x - 4)$ 14. $4(3 - 2y)$ 15. $-x(6y - 5)$

Note that the distributive property works "both directions" (we usually call one direction *distribution* and the other direction *factoring* – but they are both applications of this equality). So just as we can rewrite $3(x + 5)$ as $3(x) + 3(5)$ or $3x + 15$, we can rewrite $3x + 15$ as $3(x + 5)$.

In Exercises #16-21, use the distributive property to rewrite the expression in factored form. For example, we can rewrite $3x + 15$ as $3(x + 5)$.

16. $4x + 36$ 17. $2x + 14$ 18. $5x - 20$

19. $3x - 27$ 20. $-2x + 30$ 21. $-6x - 24$

Evaluating Expressions
Use this section prior to the module or with/after Investigation 2.

Given that $a = 2$, $b = -3$, $c = 0$, and $d = -5$, evaluate each of the expressions in Exercises #22-24.

22. $b(2a - 3d)$

23. $d^2 - 25 + c$

24. $6 - \dfrac{b}{ad}$

Given that $r = 3$, $t = 5$, $u = -2$, and $w = -12$, evaluate each of the expressions in Exercises #25-27.

25. $2(rt - w)$

26. $\dfrac{w}{3u} - r$

27. $r^2 + u^3 - t$

Formulas and Equations
Use this section prior to the module or with/after Investigation 3.

Formulas show how the values of two (or more quantities) are related as they change together. For example, the formula $A = lw$ represents how the area of a rectangle (in some square unit) is related to its length and width (measured in some units). Or, $y = 2x - 8$ represents how the values of x and y are related as they change.

Evaluating a formula involves substituting value(s) into the expression on one side of the formula (the side that contains operations and perhaps multiple terms).

In Exercises #28-30, evaluate each formula for the given input.

28. $y = (x + 5)^2 - 6$ for $x = -1$

29. $c = \frac{1}{3}(d - 4) + 7$ for $d = 12$

30. $y = \dfrac{1}{x} + x^2$ for $x = 10$

Solving an equation involves fixing the value of one side of the formula (usually the side that only contains one variable) and determining the value of the other variable(s) that produce that value.

In Exercises #31-33, use the given information to write an equation and then solve it.

31. solve $4x + 15 = y$ for x if $y = 38$

32. solve $r = 2(p - 3) + 5$ for p if $r = 17$

33. solve $z = \dfrac{x + 3}{x}$ for x if $z = 7$

Writing a formula involves representing a relationship between two (or more) quantities that change together using mathematical expressions connected by an equal sign.

34. a. Indicate on the figure and describe the attribute of the rectangle below that represents the following quantities.

    l = length of the side of the rectangle (measured in inches)
    w = width of the side of the rectangle (measured in inches)
    p = perimeter of the rectangle (measured in inches)
    A = area of the rectangle (measured in square inches)

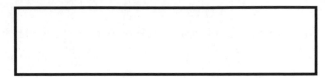

   a. Write the formula that determines the perimeter of a rectangle, given its width and length.

   b. Determine the width w of a rectangle when $p = 21$ and $l = 7$. Show your work.

35. a. Describe the attribute of the cube below that represents the following quantities.

    x = length of the sides of the cube (measured in centimeters)
    S = surface area of the cube measured in square centimeters
    V = volume of the cube measured in cubic centimeters

   b. Write the formula to express the volume of a cube V in terms of its side length x. Then determine the volume of a cube that has a side length of $4\frac{5}{8}$ centimeters.

   c. Write the formula to express the surface area of a cube S in terms of its side length x. Then determine the surface area of a cube when the cube's side length is $2\frac{2}{3}$ centimeters.

This lesson engages you in understanding ideas of quantity and variable, two foundational ideas for developing formulas and graphs to represent how two quantities change together (number of days since 3/01/2020 and the total number of COVID-19 cases in the US).

The most critical step in constructing formulas and graphs in applied problems is identifying words that provide information about a quantity—an attribute of some situation or object that can be measured. Next, representing these quantities in a diagram can help you recognize how to define one quantity in a situation "in terms of" another using a table, formula or graph. Learning to conceptualize and relate the quantities in a problem will unlock your potential to become a powerful mathematical thinker and assure your success in this and future courses in mathematics!

As we work toward the goal to relate two varying quantities with a formula or graph, it is helpful to represent the quantities in a drawing that illustrates how quantities in the problem context are related.

*1. Arizona State University graduate Desiree Linden won the Boston marathon in 2018. She began the 26.2-mile race with a slow pace due to the wind and rain. After completing the first half of the race, she increased her pace, completing the marathon in a time of 2:39:54 (2 hrs., 39 min. and 54 sec.).

   a. Identify at least two distances in this situation that are constant (their value does not vary). *Be sure to specify the value, the unit of measure, and where you are measuring from.*

   b. Identify at least two distances in this situation that are varying (their value varies). *Be sure to specify the unit of measure and where you are measuring from.*

   c. When Desiree has run 6 miles from the starting line of the 26.2-mile race, how far is she from the finish line? Indicate Desiree's approximate position on this illustration.

   d. When Desiree has run 22 miles from the starting line of the 26.2-mile race, how far is she from the finish line? Indicate Desiree's approximate position on this illustration.

   e. As the number of miles Desiree has run since starting the race increases, how does her distance from the finish line change?

---

f. Represent each of the following below the given illustration.
    i)   the value of the total length of the race using a solid line with bars at each end.
    ii) Desiree's distance from the starting line, using a dashed line with an arrow on the right end.
    iii) Desiree's distance from the finish line, using a dashed line with an arrow on the right end.

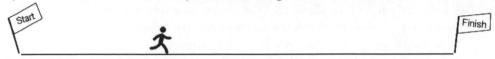

g. As Desiree runs the race, how do the lengths of the lines you have drawn in part (f) change?

h. In mathematics it is useful to use a letter (referred to as a ***variable***) to represent the value of a varying quantity. If we let the variable, x, represent Desiree's distance from the starting line (in miles),

    i) what does $26.2 - x$ represent?

    ii) what are the units?

i. Label each line that you drew in part (f) with one of the terms, 26.2, x, or $26.2 - x$.

j. What values can x assume (take on) in this situation?

k. T or F: The value of $26.2 - x$ varies as Desiree runs the race.
    T or F: The expression $26.2 - x$ represents Desiree's distance from the finish line ***in terms of*** her distance from the starting line, x. Explain what ***in terms of*** means in this situation.

l. As the value of x, Desiree's distance from the starting line (in miles) increases from 0 to 26.2, her distance from the finish line <increases or decreases> from _____
    to _____. (Be sure to include the units in your answer.)

Quantities

Quantities are the attributes of an object or situation that you can imagine measuring. To clearly describe a quantity, we must include:
*   the object or description of the situation and what attribute we are measuring
*   where we are measuring the quantity from
*   the direction of the measurement
*   the units used in the measurement

If the value of a quantity does not change then the quantity is called a ***fixed quantity*** and its value is a constant.

If the value of the quantity changes then the quantity is called a ***varying quantity*** and the quantity can assume more than one value.

---

*2 In Exercise #1, we related two varying quantities: *Desiree's distance (in miles) from the finish line of the race* with *Desiree's distance (in miles) from the starting line of the 26.2-mile race.*

a. Discuss with your classmates how to identify the quantities in a problem statement or situation.

b. What must you include in your description of a quantity so another person listening to your description is able to imagine that quantity and know how to measure it?

Variables

Instead of representing a quantity's values with words it is more concise to **define a variable** by designating some letter (or other symbol) to represent the values that a specific varying quantity can assume.

When defining a variable we must be precise in describing the quantity and how it is being measured, by saying:
  i. what quantity we are measuring
  ii. where we are measuring the quantity from
  iii. the direction of the measurement (above the ground, east of the stop sign), and
  iv. the measurement unit (inches, seconds).

Using Exercise #1, let the variable x represent Desiree's varying number of miles from the starting line of the race. In doing so, we write:

Let x represent Desiree's distance (in miles) from the starting line.

We then think of x as having a value that changes as Desiree runs the race so that it always represents her distance (in miles) from the starting line.

Example of *a varying quantity:* The height (in feet) of an airplane above 5000 feet elevation.
In this situation, the airplane's *height* is a measurable attribute of the airplane. The airplane's height is being measured from 5000 feet above sea level. The unit of measurement is feet.

Example of a *fixed quantity*: The height of the Washington Monument (in feet).
In this situation, the Washington monument's *height* is a measurable attribute of the Washington monument. The height is being measured from the ground on which the monument sits. The unit is feet.

---

3. A tortoise and hare are competing in a race around a 400-meter track. The arrogant hare decides to let the tortoise have a 150-meter head start. When the starting gun is fired, the hare begins running at a constant speed of 6.5 meters per second and the tortoise begins crawling at constant speed of 2.5 meters per second.

a. Underline each phrase in the above text that provides information about a quantity. Use a solid line for phrases conveying information on fixed quantities and a dashed line when underlying phrases that provide information on varying quantities.

b. List the *varying* quantities that you have identified. *Be sure to say where the quantity is being measured from, the direction of measurement, and the unit of measurement.*

c. Illustrate the distances described in this situation in a drawing. Use a solid line with bars on the end to represent a fixed distance, and a dashed line with an arrow on one end to represent any varying distances.

d. Define a variable to represent the number of seconds since the starting gun was fired. Be specific in your description, make sure you address criteria (i) – (iv) when defining a variable.

e. Use the variable you defined in part (d) to represent the hare's distance from the starting line.

f. Use the variable you defined in part (d) to represent the tortoise's distance from the starting line.

g. Use the variable you defined in part (d) to represent the hare's distance from the finish line.

h. Use the variable you defined in part (d) to represent the tortoise's distance from the finish line.

4a. Identify and describe two fixed and two varying quantities in your classroom by describing a measurable attribute of some object. *Be sure to say where the quantity is being measured from, the direction of measurement, and the unit of measurement.*

*b. Determine if each of the following describes a quantity. If not, modify the statement so that it does describe a quantity.
  i. Cases of COVID-19.

  ii. The height of a ball.

  iii. The distance Bella has walked from a stop sign.

*5. A ball is thrown upward by a person standing on the ground. If we want to represent *the height of the ball above the top of a 15-foot tall building,* we can assume that the building's height is being measured from the ground and that the ball can be both above and below the top of the 15-foot tall building.

Here is a number line with 0 representing the top of the 15-foot tall building. We labeled the number line so that numbers to the right of 0 represent the ball's height **above** the top of the building (number of feet).

the ball's height above the top of the building (number of feet)

a. What do the negative values indicate in this situation?

b. Define a variable d to represent the height (in feet) of the ball above the top of the 15-foot tall building.

c. What does a value of $d = -5$ convey about the height of the ball relative to the top of the building?

d. Illustrate on the given number line the ball's height above the 15-foot building increasing from $d = -6$ feet to $d = 8$ feet. How many feet did the ball travel, and in what direction was the ball traveling?

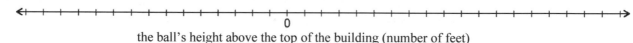

0

the ball's height above the top of the building (number of feet)

Note that in each description of a varying quantity you must have a starting point for your measurement in mind. "The distance of a car in miles" does not describe a quantity, but "The distance traveled in miles since the car left the parking lot" does.

*6. We can represent the distance (in miles) of Laura's car north of a stop sign using a number line.

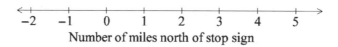

Number of miles north of stop sign

a. What does 0 on the number line represent?

b. Laura is 2 miles south of the stop sign. Illustrate this on the number line.

c. Using a dashed line with an arrow, illustrate Laura's distance from the stop sign as she drives from the position 2 miles south of the stop sign to a position 4 miles directly north of the stop sign.

*7. a. Discuss the usefulness of a variable (symbol) for representing the values of a varying quantity.

b. It is important when defining variables that we are specific in describing the varying quantity whose values are being represented by the variable (designated symbol).

Is the statement, "d = Erin's distance (in miles)" an adequate definition of a variable? If not, what is missing and how might you modify this statement so that it defines a variable?

c. When defining a variable, why is it important to describe where the quantity is being measured from, the direction of the measurement, <u>and</u> the measurement unit?

*8. Jenny's gas tank has 3.5 gallons of gasoline remaining when she pulls into a gas station to fill her 25-gallon tank.
  a. Represent the quantities in this situation by:
    i. drawing a solid line to represent the fixed quantity of 25 gallons the tank will hold.
    i. drawing a solid line to represent the fixed quantity of 3.5 gallons.
    ii. drawing a dashed line with an arrow to represent the varying number of gallons of gasoline. that Jenny can add to the tank.

  b. If n represents the varying number of gallons of gasoline Jenny can add to her tank, write an expression to represent the varying number of gallons of gasoline in the tank.

9. The city's water tank holds 20,500 gallons of water. After the tank is full, water begins draining from the tank.
  a. Define a variable d to represent the number of gallons of water that has drained from the tank.

  b. Use the variable d, and the fixed quantity of 20,500 gallons the tank can hold, to write an expression that represents the number of gallons of water remaining in the tank.

  c. What values can d assume (take on) in this situation?

  d. As the number of gallons of water that have drained from the tank d increases from 0 to 20,500 how does the number of gallons of water remaining in the tank change?

This lesson will improve your ability to represent how two quantities change together using formulas and graphs. In addition to conceptualizing the quantities in a problem context and defining variables to represent the quantities' values, it is important to understand what operations to use to combine specific quantities described in an applied problem. We begin the lesson by exploring a key question:

What do the results of performing the operations addition, subtraction, multiplication, and division represent?

*1. If the variable x represents the total number of COVID-19 deaths in the United States since March 1, 2020, what do the following expressions represent?

a. $x - 100,000$

b. $1.2 x$

c. $\dfrac{x}{1,000}$

*2. If d_t represents the number of feet a tortoise has crawled north of the starting line of a race and d_h represents the number of feet a hare has traveled north of the same starting line, what do each of the following expressions represent?

a. $d_t - d_h$

b. $\dfrac{d_t}{12}$

c. $\dfrac{d_h}{d_t}$

*3. The formula $C = 2\pi r$ is a formula that defines how a circle's circumference C is related to the circle's radius r (both measured in the same length units). Complete the following statements.

a. The circumference of any circle is _____ times as large as the circle's radius.

b. The radius of any circle is _____ times as large as the circle's circumference.

c. Describe at least two different ways to interpret what $\dfrac{C}{2\pi}$ represents.

d. Describe at least two different ways to interpret what $\dfrac{C}{r}$ represents.

e. Given the radius of a circle (drawn below), construct a line segment with a length approximately equal to the circle's circumference. How are the lengths of these segments related?

_____

*4. Becky and Wendy are 27 feet apart and start walking toward one another at the same time. Wendy is walking twice as fast as Becky, so when Becky travels x feet, Wendy travels $2x$ feet.

a. Illustrate this situation with a drawing, given that x represents the distance Becky has walked since Becky and Wendy started walking toward one another.
(*Use a solid line to represent the fixed distance between Becky and Wendy, and dashed lines with arrows (vectors) to represent the varying distances that Becky and Wendy have walked.*)

b. Label each line segment in your diagram in part (a) with one of the expressions, x, $2x$ or $27 - 3x$, that represents its value. Then describe what each expression represents.

c. If we let d represent the varying distance (in feet) between Becky and Wendy, then $d = 27 - 3x$. What is the benefit of representing the varying distance between Becky and Wendy using the single variable d?

d. Brian drew a line segment on which he will represent the number of feet (x) Becky has walked toward Wendy. He drew a second line segment on which he will represent the number of feet (d) between Becky and Wendy.

   i)   Represent each of the values $x = 0$, $x = 3$, $x = 7$, $x = 9$ with a point on the bottom line segment.

   ii) Determine the corresponding values of d, given that $d = 27 - 3x$, then plot each value with a point on the top line segment.

   iii) Connect each corresponding value of x and d with a straight line.
        (Begin by connecting $x = 0$ to $d = 27$.).

e. Discuss how to configure the two line segments in part (d) so that you can represent each corresponding value of x and $27 - 3x$ with a single point. Redraw the line segments so it is possible to represent the 4 corresponding values of Becky's distance walked, x, and the distance between Wendy and Becky, d, with 4 distinct points.

A ***coordinate plane*** consists of a vertical axis and horizontal axis that intersect at a point. The point of intersection of the two axes is called the ***origin*** and is labeled (0, 0).

It is a mathematical convention to let the horizontal axis represent the values of the independent quantity and the vertical axis to represent the values of the dependent quantity. The dependent quantity is typically defined ***in terms of*** (or using the variable that represents the value of) the independent quantity.

*5. Juan and Brandon are 45 feet apart and start walking toward one another at the same time. Juan is walking twice as fast as Brandon, so when Brandon travels x feet, Juan travels $2x$ feet. We wish to define the distance (in feet) between Juan and Brandon since they started walking ***in terms of***, x, the number of feet Brandon has walked toward Juan.

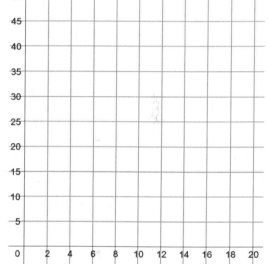

    a.   Which axis should be labeled, "the distance in feet Brandon has traveled since he started walking toward Juan"?

    b.   Which axis should be labeled, "the distance in feet between Brandon and Juan since they started walking toward one another"?

    c.   Write an expression that represents the distance between Juan and Brandon ***in terms of*** x, the number of feet Brandon has walked toward Juan.

    d.   Label the axes and plot the points (0, 45), (10, 15) and (15, 0) on this grid. What does the point (10, 15) represent in this context?

    e.   When $x = 5.2$, what is the value of $45 - 3x$? Plot the point on the grid.

    f.   Draw a straight line through the points you plotted on the grid. What do the points that make up this line represent?

    g.   When the distance Brandon has walked increases from 0 to 5 feet, how does the distance between Brandon and Juan vary?

    h.   Define a formula that represents the distance in feet between Brandon and Juan, d, in terms of the number of feet, x, that Brandon has walked toward Juan.

*6. A 7-quart glass container weighs 4 pounds when empty. The weight of water is 2 pounds per quart. Let x represent the varying number of quarts of water that are in the container.

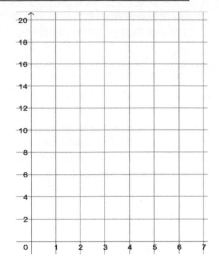

a. Label the axes on the grid and plot the points (0, 4), (1, 6) and (4, 12). What do these points represent in this context?

b. Draw a straight line through these points. What do the points that make up this line represent?

c. Write a formula to represent the total weight of the container and water, y, (in pounds) in terms of the number of quarts of water that have been added to the container, x.

Changes in Quantities

Once we have defined a variable x to represent the varying values of a quantity, we can write Δx (read as "change in x") to represent **changes** in that quantity's value.

The change in a quantity's value is a new quantity itself. If the value of x changes from x_1 to x_2, then the amount of change can be represented as $x_2 - x_1$.

For example, if x changes from $x = 6$ to $x = 19$, then $\Delta x = 19 - 6$, or the value of x changed by 13 units.

---

7. The temperature in Phoenix, T, in degrees Fahrenheit varies throughout the day. On May 1st the temperature was 65 degrees Fahrenheit at 6 am and 90 degrees Fahrenheit at 11 am.

a. What was the change in temperature, ΔT, over the time period from 6 am to 11 am?

b. Represent the change in temperature, ΔT, determined in part (a) on the number line.

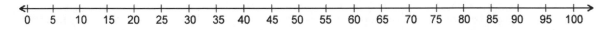

c. Represent a change of temperature of 5 degrees Fahrenheit as the temperature increases from 65 degrees Fahrenheit to 90 degrees Fahrenheit on May 1st. How many ways can you represent this 5-degree change?

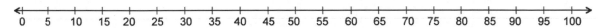

d. Represent a change of −25 degrees Fahrenheit from 80 degrees Fahrenheit.

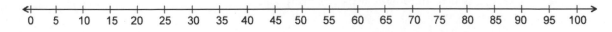

---

*8. Sam boards a Ferris wheel from the bottom and rides around several times before getting off. This graph represents Sam's height above the ground (in feet) with respect to the amount of time (in seconds) since the Ferris wheel began moving for one complete rotation of the wheel.

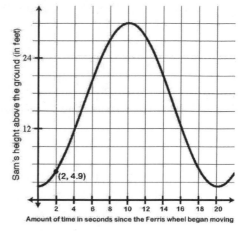

a. The point (2, 4.9) is plotted on the graph. Explain what this point conveys in the context of the Ferris wheel situation.

b. Complete the sentence, "As the number of seconds since the Ferris wheel began moving increases from 0 to 20 seconds, Sam's height above the ground…

c. As the *number of seconds since the Ferris wheel began moving* **increases** from 2 to 8 seconds, how does *Sam's height in feet above the ground* **change**? Represent this change in height on the graph.

d. As the *number of seconds since the Ferris wheel began moving* **increased** from 10 to 16 seconds, what was the **change** in *Sam's height in feet above the ground*? Represent this change on the graph and with a mathematical expression.

c. Let *t* represent the values on the horizontal axes. Write a complete sentence that describes what the variable *t* represents in the context of this situation.

Let *t* represent……

f. Define the variable *h* so that it represents the varying values of the dependent quantity in this situation. *When defining variables remember that you must ALWAYS say where the measurement begins (for example, "since 10 am"), the direction of the measurement (north of the starting line), and state the units (seconds, feet, square inches, etc.).*

Let *h* represent…

9. The given graph represents the depth of water in a reservoir in terms of the number of months since January 1, 1990.

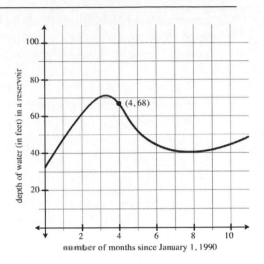

a. Define variables to represent the values of the varying quantities in this situation.

b. Interpret the meaning of the point (4, 68).

c. Label 2 other points on the graph and describe what they represent.

d. Label the point (6, 20) on the grid. What does this point represent?

e. The number of months since January 1, 1990 increases from 1 to 3 months.
   i. What is the change in the number of months since January 1, 1990?

   ii. Represent this change on the graph.

f. As the number of months since January 1, 1990 increases from 1 to 3 months…

   i. How does the depth of the water in the reservoir vary?

   ii. What is the change in the depth of the water?

   iii. Represent this change of the water depth of the reservoir on the graph.

g. Estimate the depth of the water in the reservoir 7 months after January 1, 1990.

h. How does the depth of the water in the reservoir vary as the number of months since January 1, 1990 increases from 4 to 7 months? Illustrate this variation on the graph.

You have been studying linear relationships and the idea of constant rate of change since your first course in algebra. You learned that any formula that can be written in the form $y = mx + b$ describes a linear relationship between x and y, and that x and y vary together at the constant rate of change, m, and b represents the *y-intercept*, or value of y when $x = 0$.

In this lesson, we will explore in more depth what it means for two quantities to change together at a constant rate of change. In particular, **when saying that x and y are changing together at a constant rate of change, m, what constraint does this impose on how x and y are changing together?**

*1. Imagine yourself riding on a stationary bike. After your 2-minute warm up, you begin riding at a constant speed of 4 minutes per mile for the next 5 miles.

    a.   What 2 quantities are changing together in this context?

    b.   As you are riding your stationary bike, the number of minutes you have been riding, t, and the number of miles you have been riding, d, are changing together. Discuss with your classmates what it means to be traveling at the constant speed of 4 minutes per mile. (*Say more than, "you travel for 4 minutes every time you travel a distance of 1 mile," or that "your speed doesn't change."*)

    c.   When riding at a constant speed of 4 minutes/mile:
        i.   How many miles would you ride in 1 minute? Explain your thinking.

        ii. How many miles would you ride in 6 minutes? Explain your thinking.

        iii. How many miles would you ride in 7 minutes? Explain your thinking.

        iv. How many miles would you ride in ½ minute? Explain your thinking.

    d.   What generalizations can you make about how your change in time, Δt, and change in distance, Δd, are constrained during the period when you are riding at a constant speed of 4 minutes per mile?

    e.   When riding at a constant speed of 4 minutes per mile:
        i.   How many minutes will it take you to ride ½ mile? Explain your thinking.

        ii. How many minutes will it take you to ride $\dfrac{1}{10}$ mile? Explain your thinking.

        iii. How many minutes will it take you to ride 3.2 miles? Explain your thinking.

f. Use the coordinate plane to plot the point (0, 2) on the axes and describe what this point represents in the context of this situation.

   i.   Use your answer to part (d) to plot 3 more points on your graph.

   ii. Draw a line through the points. What do the points that make up this line represent?

   iii. Illustrate on your graph an increase in *d* from 1 to 4 miles. Determine the corresponding value of Δ*t* and illustrate this change on your graph.

   iv. Complete this sentence: When riding at a constant speed of 4 minutes/mile, any change of 1.5 miles will result in a change of _____ minutes.

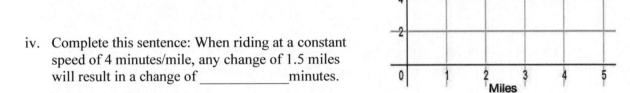

   v.   Construct a formula to represent the change in riding time, Δ*t*, in terms of the change in the number of miles ridden, Δ*d*. (Hint: A change in your riding time is always how many times as large as a change in your distance traveled?)

g. Construct a formula to represent the total number of minutes you have been riding, *t*, in terms of the number of miles, *d*, since you started riding at a constant rate of 4 minutes/per mile. Which variable is the independent variable in this formula?

2. Benny walks to school at a constant speed and travels 8 feet in 3 seconds.
   a.   If Benny walks for 1 second how far will he walk? Explain.

   b.   If Benny walks for 4 seconds how far will he walk? Explain.

   c.   T or F: To say that Benny walks at a constant speed and travels 8 feet in 3 seconds is the same as saying that Benny walks at a constant speed of $\frac{8}{3}$ feet/second. Explain.

   d.   If Benny continues at the same constant rate of change for 6.5 seconds after reaching school, what distance will he travel in those 6.5 seconds? Show work.

Constant Rate of Change

For any pairs of values (x_1, y_1) and (x_2, y_2) in the relationship, the ratio of $y_2 - y_1$, (read as the change in y) to $x_2 - x_1$ (read as the change in x) remains constant (or always has the constant value **m**). We can also say that if two quantities are related by a constant rate of change, the corresponding changes in the two quantities are proportional.

Using this definition, if we let $\Delta x = x_2 - x_1$ and $\Delta y = y_2 - y_1$, for any pairs of values (x_1, y_1) and (x_2, y_2) in the constant rate of change relationship between x and y, then we can also write $\dfrac{\Delta y}{\Delta x} = m$.

---

*3. The owner of a swimming pool wants to know the rate at which his swimming pool is being filled. When he begins filling the pool at 2 p.m. the depth of the water is 18 inches. Water flows in at a constant rate of change.

    a. When it is 5 p.m. the depth of the water is 29 inches. What is the change in the depth of water in the pool over the time period from 2 p.m. to 5 p.m.?

    b. At what rate of change is the pool being filled? (*Be sure to include the units*.)

    c. If the pool continues to be filled for another 4.5 hours (after 5 p.m.),
       i. what is the change in the depth of the water in the pool?

       ii. what is the depth of the water in the pool?

4. Porter received a remote-control car as a gift for his birthday. Porter's car began traveling at top speed 6 meters after it began to move and continued at this speed for 5 more seconds. When traveling at top speed his car traveled 7 meters every 2 seconds.

    a. What was the top speed of Porter's car?

    b. Define the distance, *d*, (in meters) Porter's car traveled *since it started to move* in terms of the number of seconds, *t*, *since Porter's car reached top speed*.

    c. How far did Porter's car travel at top speed over any:
       i. ½ second interval after it reached top speed?

       ii. 3 second interval after it reached top speed?

       iii. 3.5 second interval after it reached top speed?

---

d. How far did Porter's car travel at top speed over any Δt second interval ($0 \le t \le 5$) after Porter's car reached top speed?

e. Construct a graph to represent the distance, d, (in meters) Porter's car traveled since it started to move, in terms of the number of seconds, t, since Porter's car reached top speed. *Be sure to label the axes.*

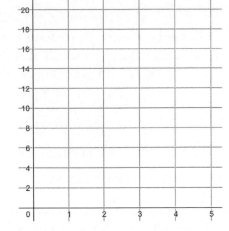

f. How is m, the constant rate that Porter's car traveled after reaching top speed, represented on the graph?

g. What does b (the vertical intercept) represent in this context?

Constant Rate of Change Revisited

If y changes at a constant rate with respect to x, then we know that $\frac{\Delta y}{\Delta x} = m$, or $\frac{y_2 - y_1}{x_2 - x_1} = m$ when (x_1, y_1) and (x_2, y_2) represent any two pairs of corresponding values in the relationship.

Note that $\Delta y = m \cdot \Delta x$ is another way to express the relationship $\frac{\Delta y}{\Delta x} = m$. This form emphasizes that Δy is always m times as large as Δx when x and y change together. Note that we can also write this as follows.

$$\Delta y = m \cdot \Delta x$$
$$y_2 - y_1 = m \cdot (x_2 - x_1)$$

---

*5. Each of the following represents a situation with a constant rate of change. Use the definition above to explain what these statements mean.

a. $\Delta y = 6 \cdot \Delta x$ (*Hint: As the values of x and y change together, the change in y is always...*)

b. $\Delta y = -2 \cdot \Delta x$

*6. a. Given that x and y are changing together at a constant rate of change so that $\Delta y = -2 \cdot \Delta x$ and when $x = 1, y = 7$, construct the graph of y in terms of x.

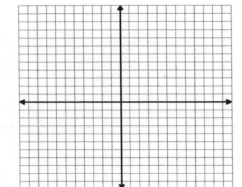

b. As x increases from 1 to 5.5, how does y change? Illustrate these changes on your graph.

c. Does x take on all values between 1 and 5.5? If so, how is this illustrated on your graph?

In your high school Algebra classes you learned that the constant rate of change of y with respect to x can be determined if you know any two points in the relationship in which x and y are changing together at a constant rate of change (also referred to as a linear relationship). We reviewed this in our definition boxes on constant rate of change, but restate here that if (x_1, y_1) and (x_2, y_2) are **any two pairs of values in a linear relationship**, then we can determine the change from x_1 to x_2 and the corresponding changes from y_1 to y_2. The ratio of these changes $\dfrac{y_2 - y_1}{x_2 - x_1} = m$ is the constant rate of change of y with respect to x.

7. The given graph represents a linear relationship between y, Bob's distance (in miles) to the finish line of a race, and x, the amount of time (in minutes) since Bob passed the final water station.

    a. What does the point $(-22, 8.2)$ represent in this situation?

    b. Determine the value of m, the constant rate of change of y with respect to x, in this linear relationship and say what this value represents.

    c. Use the general formula $y_2 - y_1 = m \cdot (x_2 - x_1)$ to relate the values of x and y in the linear relationship depicted in the given graph.

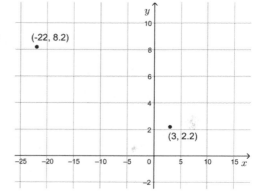

    d. For the point (3, 2.2) we have that $x_1 = 3$ and $y_1 = 2.2$.

        i. Since y represents the varying distances that Bob is from the finish line of the race, what does $y - 2.2$ represent?

        ii. Since x represents the varying number of minutes since Bob passed the final water station in the race, what docs $x - 3$ represent in this situation?

Point-Slope Form of a Linear Function

The general formula $y - y_1 = m \cdot (x - x_1)$ defines a linear relationship where m is the constant rate of change of y with respect to x and (x_1, y_1) is an ordered pair in the relationship.

8. a. What do x and y represent in thc point-slope form of a linear function?

    b. Describe situations in which the point-slope form of a linear function is useful.

---

*9. Given that the x and y axes have the same scale for both of these graphs, answer the following questions.

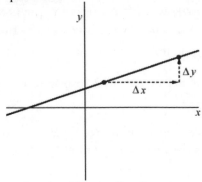

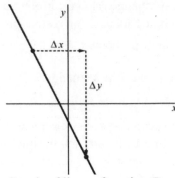

Graph of linear function A   Graph of linear function B

a. What is the approximate constant rate of change of y with respect to x for the linear functions?

Linear function A: Linear function B:

b. Write a formula that expresses Δy in terms of Δx .
Linear function A: Linear function B:

c. Write a formula that expresses y in terms of x given the y-intercept of the graph of function A is 3 and the y-intercept of the graph of function B is –3.

Formula for linear function A: Formula for linear function B:

10. A candle is 8.5 inches tall when it has been burning for 6 hours at a rate of ¼ in. per hour.

a. How tall was the candle before it was lit (when it had been burning for 0 hours)?

b. How tall was the candle when it had been burning for 2.4 hours? (*Hint: Think about how the change in the remaining length of the candle is related to the change in the number of hours the candle has been burning.*)

This investigation examines the special case of constant rate of change situations where the following two facts are true.

- There exists a constant rate of change (changes in the two quantities are proportional).
- The quantities themselves are proportional.

Recall that two quantities are said to be proportional if their values, *x* and *y*, are related by a constant multiple, *m*, <u>as they change together</u>. That is, $y = mx$ for a constant *m*, or equivalently the ratio of their values is always *m* ($y/x = m$ for the same constant *m*).

---

*1. A photographer has an original photo that is 6 inches high and 10 inches wide and wants to make different-sized copies of the photo so that the new photos are not distorted. (Assume this constraint for all questions in this exercise.)

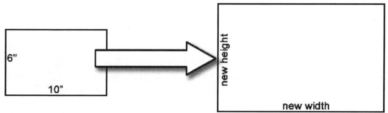

a. The photographer wants to enlarge the original photo so that the new photo has a width that is 25 inches, what will the height of the new photo need to be so that the image is not distorted? Explain the reasoning you used to determine your answer.

b. What ratio must remain constant to assure that the photo is not distorted as it is resized?

c. Create a table that gives the heights of resized photos corresponding to the new desired photo widths of 1 inch, 2 inches, 5 inches, and 25 inches. Discuss with your classmates (or give a written explanation for) why the dimensions you found will not distort the photo.

Resized Photo width (inches)	Resized Photo height (inches)
1	
2	
5	
25	

d. Construct a graph of the height vs. width to illustrate the relationship between the height of the resized photo for any choice of width for a photo. Be sure to label your axes with the appropriate quantities and units, with the photo widths represented on the horizontal axes.

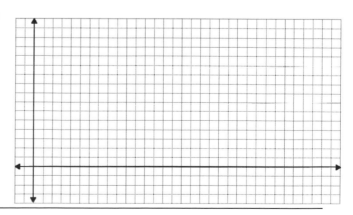

e. Explain what a point on your graph represents.

f. Define variables to represent values of the resized photo's height, h, and width, w. Then write a formula to determine the height of a resized photo given (in terms of) the width of a resized photo (so that the resulting photo is not distorted).

g. Fill in the blank: The height of a resized photo is always _____ times as large as the width of a resized photo, given that the original photo is 10" wide and 6" high.

h. Fill in the blank: The ratio of the width to the height of a resized photo is always _____.

i. Illustrate the following on the graph you constructed in part (d).

    i. The point that represents a 5-inch wide and 3-inch high photo on your graph.

    ii. A line segment that represents an increase of 2.5 inches in the photo's width (from a 10-inch wide photo), and a line segment that represents the corresponding change in the photo's height.

j. Will a change of 2.5 inches in width always correspond to a change of 1.5 inches in height? Why or why not? How can you see this on the graph?

k. If the change of the width of the photo is three times as large as a previous change in width, will the change in height also be three times as large as the previous change in height? Explain.

l. Use the definition of proportionality (at the beginning of this investigation) to determine if the resized photo's height is proportional to the resized photo's width. Discuss with your classmates and justify your answer.

m. Fill in the blank: For any change in the resized photo's width, Δw, the change in the resized photo's height, Δh, is always _____ times as large. How can you see this on your graph?

*2. Here are two cylinders: one wide and one narrow. Both cylinders have equally spaced marks for measurement. Water is poured into the wide cylinder up to the 4th mark (see A). This water rises to the 6th mark when poured into the narrow cylinder (see B).

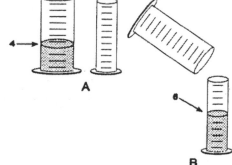

a. Both cylinders are emptied, and water is poured into the narrow cylinder up to the 11th mark. How high would this water rise if it were poured into the empty wide cylinder?

b. Imagine pouring water into the narrow cylinder and let x represent the height (mark-number) the water reaches in the narrow cylinder as you pour. Let y represent the height (mark-number) the same volume of water would reach if poured into the wide cylinder.

   i.   What is the meaning of $\frac{y}{x} = \frac{2}{3}$ in this context?

   ii. What is the meaning of $y = \frac{2}{3}x$ in this context?

   iii. What is the meaning of $\Delta y = \frac{2}{3} \cdot \Delta x$ in this context?

c. i.   When $x = 3$, what is the value of y?

   ii. When $x = 4.5$, what is the value of y?

d. Compare and contrast what the three statements, $\frac{y}{x} = \frac{2}{3}$, $y = \frac{2}{3}x$, and $\Delta y = \frac{2}{3} \cdot \Delta x$, convey about how the water heights and changes in water heights for the two cylinders are constrained and related.

Proportional Relationships

If y varies with x in a linear relationship so that $y = m \cdot x$ (that is, the value of y is m times as large as the value of x), then we can think of any value of x or y as a change away from 0.

In this instance, we say that y **is proportional to** x (and also that x is proportional to y).

Note that if $y = m \cdot x$, then it's also true that $\Delta y = m \cdot \Delta x$ (that is, the change in y is m times as large as the change in x as the quantities co-vary). We can say that Δy is proportional to Δx and that Δx is proportional to Δy.

---

3. The price of ground beef is $4.50 per pound.

   a. Define an ***expression*** to represent the cost of purchasing n lbs. of ground beef.

   b. Define a ***formula*** to represent the cost, c, (in dollars) of n pounds of ground beef.

   c. Construct a graph to represent the cost, c, (in dollars) of n pounds of ground beef. (Label your axes.)

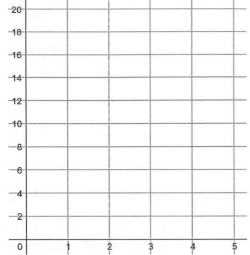

   d. Discuss with your classmates how you could justify that the number of pounds of ground beef purchased, n, is proportional to the cost, c, of this purchase. Provide at least two different written justifications.

   e. If you have 2 pounds of ground beef on the scale, and then add another 1.5 pounds of ground beef to the scale, what is the change in the cost of the purchase, Δc?

   f. Will an increase of 1.5 pounds of ground beef always result in the same increase in cost? Discuss and provide at least two different justifications for your answer.

   g. T or F: If n and c are proportional, then n and c are changing together at a constant rate of change. (Justify your answer or provide a counter example.)

   h. T or F: If n and c are changing together at a constant rate of change, then n and c are proportional. (Justify your answer or provide a counter example.)

4. A 19-inch candle is lit. The candle burns away at a constant rate of 2.5 inches per hour.
   a. Draw a diagram that represents a "snapshot" of the candle as it is burning. Represent the fixed quantities with a solid line, and each varying quantity with a dashed line and arrow.

   b. Define an *expression* to represent the remaining length of the candle in terms of the number of hours, t, since the candle started burning. Label the line segment representing this length in your diagram in part (a) with this expression.

   c. Define a *formula* to represent remaining length of the candle, r, (in inches) in terms of the number of hours, t, since the candle started burning.

   d. Represent the change in the remaining length of the candle (in inches), Δr, in terms of the change in the number of hours, Δt, the candle has been burning.

   e. What quantities in this situation are proportional? Explain.

   f. T or F: Since the change in the remaining length of the candle, Δr, is proportional to the change in the number of hours the candle has been burning, Δt, the remaining length of the candle (in inches), r, is proportional to the number of hours the candle has been burning since it was first lit, t. Justify your answer.

   g. How much time will it take for the candle's length to become 0 (to burn the entire candle)? Show your work and explain your reasoning using the proportional relationship that constrains how Δr and Δt change together.

   h. Explain in your own words how the idea of *proportional quantities* is related to and different from the idea of *constant rate of change of one quantity with respect to another*.

*5. A 17-inch candle is lit and burns at a constant rate of 1.8 inches per hour. Let t represent the number of hours since the candle started burning. Let L represent the remaining length of the candle.

a. What does –1.8 represent? What are the units and what do they represent?

b. What does the expression $1.8t$ represent in this context?

c. What does the expression $17 - 1.8t$ represent in this context?

d. Is ΔL proportional to Δt? Justify your answer.

e. Is L proportional to t? Justify your answer.

f. How are variables, expressions, and formulas related and different?

*1. a. A policeman is positioned on a side-road *without* a radar gun. Is it possible for the officer to determine the speed of a car as it passes the officer? Explain.

    b. If the policeman had a radar gun, would it be possible for the policemen to determine the instantaneous speed of the car? (Consider what a radar gun measures to determine a car's speed.) Support your answer with an explanation.

*2. A car is driving away from a crosswalk. The distance, *d*, (in feet) of the car north of the crosswalk *t* seconds since the car started moving is given by the formula $d = t^2 + 3.5$.

    a. As the number of seconds since the car started moving increases from 1 second to 3 seconds, what is the change in the car's distance north of the crosswalk? (Use the given formula to determine this change in distance.)

    b. Label the axes and construct a graph that represents the distance (in feet) of the car north of the crosswalk, *d*, in terms of *t*, the number of seconds since the car started moving. Plot the points: (0, 3.5) (1, 4.5), etc. and sketch the curve.

    c. What do the points that make up the curve represent?

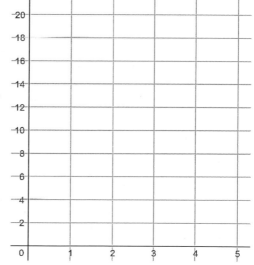

    d. Illustrate each of the following on your graph.

        i. An increase in *t* from 1 to 3 seconds.

        ii. The corresponding change in the car's distance north of the crosswalk, as the value of *t* increased from 1 to 3 seconds.

    e. True or False: The car travels at a constant speed as the value of *t* increases from 1 to 3 seconds. Discuss and explain. (*Hint: It may help to think about how far the car travels in the 1ˢᵗ second as compared to how far it travels in the 2ⁿᵈ second*).

f. Since the car is not traveling at a constant speed, it can be challenging to describe the car's speed. It is a common practice to <u>estimate</u> the speed of an object over an interval of time by pretending its speed is constant, and asking, "At what constant speed would the object have to travel in order to travel the same distance in the same amount of time?"

For this problem context, answer the following questions to determine the constant speed needed to travel the same distance that the car actually traveled over the time interval $t = 1$ to $t = 3$ seconds.

   i. Plot the points (1, 4.5) and (3, 12.5) on the graph you created in part (b), and draw a line passing through them.

   ii. Determine the constant rate of change of the linear relationship (slope of the line) represented by the line you drew. (Be sure to include units in your answer.)

g. True or False: The points making up the line you drew represent the actual distance (in feet) of the car north of the crosswalk for various values of t, as t increases from 1 to 3 seconds. Explain.

h. Using the formula $d = t^2 + 3.5$, what is the constant speed the car would need to travel on the interval from $t = 3$ seconds to $t = 4.5$ seconds to achieve the same change in distance that the car actually traveled. (Be sure to include units in your answer.)

*3 a. The distance, d, (in feet) of a car south of an intersection t seconds after it started moving is given by the formula $d = 2t^2 - 8t + 8$.
   [*Note: We will use subscripts to refer to different values of d and t.*]

b. Determine the value of d_1 when $t_1 = 2$, and describe what the value conveys about the car's distance south of the intersection.

c. Determine the value of d_2 when $t_2 = 5$, and describe what the value conveys.

d. Determine the value of $\frac{d_2 - d_1}{t_2 - t_1}$ and explain what this value represents in the context of this situation (*Hint: What constant rate of change is this?*)

e. Label the axes and construct a graph of
$d = 2t^2 - 8t + 8$ for values of $t \geq 0$.
(You may use a graphing calculator or
set up a table.)

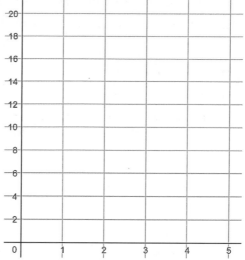

f. Describe how the car's distance south of the
intersection varies over the first 5 seconds since
it started moving.

g. On your graph, illustrate the constant speed another car would need to travel in order to
cover the same distance (as the car actually traveled) over the time interval from $t_1 = 2$ to $t_2 = 5$.

h. If the car continued traveling at a constant speed of 12 feet per second on the interval from $t = 5$
to $t = 11$ seconds since the car started moving, how far would the car be south of the stop sign 11
seconds after it started moving?

Average Rate of Change (over some interval)

The *average rate of change* of a function over some
interval of the domain is the constant rate of change over
that interval that produces the same net change in the
function's output value.

Visually, it's the slope of the line passing through the
points at the beginning and end of the interval, so it makes
sense that the average rate of change is calculated using the
slope formula $m = \frac{y_2 - y_1}{x_2 - x_1}$.

The average rate of change is a common tool for describing
the general behavior of a function over some interval when
the rate of change is not constant.

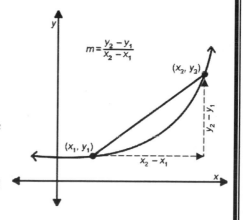

You have likely observed that determining an average speed of some moving object on a time interval from t_1 to t_2 involves determining the constant speed needed to travel the same distance that the object actually traveled over the specified time interval (from t_1 to t_2).

*4. Given that a car's distance (in feet) east of an intersection t seconds since it started moving is represented by $d = t^2 - 2t$, determine the car's average speed over each interval of elapsed time. *Be sure to include the units in your answer!*

    a. $t_1 = 2$ to $t_2 = 6$ b. $t_1 = 6$ to $t_2 = 10$ c. $t_1 = 10$ to $t_2 = 11$

    d. What observations can you make about how the car's speed is changing on the time interval from $t_1 = 2$ to $t_2 = 11$ since the car started moving east of the intersection.

5. Given that a car's distance (in feet) north of an intersection t seconds since it started moving is represented by $d = (t-4)(t+2)$, determine the car's average speed over each interval of elapsed time. *Be sure to include the units in your answer!*

    a. $t_1 = 1$ to $t_2 = 5$ b. $t_1 = 3$ to $t_2 = 5$ c. $t_1 = 4.5$ to $t_2 = 5$

    d. What observations can you make about how the car's speed is changing on the interval from $t_1 = 1$ to $t_2 = 5$?

*1. You are considering buying a house located at the intersection of D Street and 23rd Avenue. Your cell phone provider is building a cell tower at the intersection of B Street and 20th Avenue. Each street is located one mile apart. If you are farther than 3.5 miles away from the cell tower, you will not have good reception.

a. What quantity do you need to find to determine if you will have good reception at the house? Illustrate these quantities on the figure below.

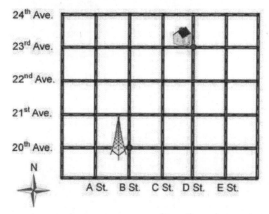

b. Determine the distance between the cell tower and the house.

c. Define variables that can be used to describe the distance of *any house* from the tower.

d. Use the variables defined in (c) above to define a formula that represents the distance of *any house* from the tower. *Hint: Use the Pythagorean theorem as a starting point.*

c. Another cell tower is located at some point (x_1, y_1). Define a formula to determine the distance of any house from that cell tower.

*2. Consider a cell tower at the location (–2, 4) on the coordinate plane with a coverage radius of 3 miles. Note that every point 3 miles away from the cell tower will be on the boundary of the cell coverage. This will form a circle around the point (–2, 4) with a radius 3 units long.

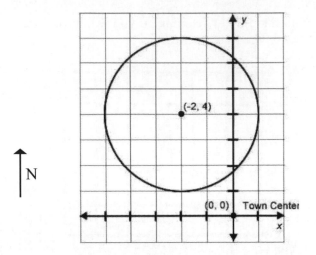

   a.   A point $(x, 2)$ southeast of the tower is on the service boundary. What must be the value of x?

   b.   Let (x, y) be any point on the service boundary. Create a formula that represents the distance of any house from the cell tower located at (–2, 4). Clearly define any variables you use.

*1. A carpenter needs to cut a wooden board to a length of 7.6 inches with a **tolerance** of 0.1 inch, meaning the actual length after the cut can be within 0.1 inch of 7.6 inches and still be usable.

    a. Let x represent the actual length of the board after cutting. List some possible values for x that represent usable board lengths, and then represent all possible values on the number line.

    b. Write a mathematical statement (inequality) to represent the possible usable board lengths.

    c. We will call the difference between the actual board length and the desired board length the ***error***. Write an expression to represent the error.

    d. Write a mathematical statement (inequality) that represents errors within our tolerance of 0.1 inch.

    e. Write a mathematical statement (inequality) that represents actual board lengths x that are not usable, then represent these values on a number line.

*2. Suppose we want to choose an x value within 4.5 units of 16.

    a. List some values of x that meet this constraint.

    b. Write a mathematical statement (inequality) to describe the possible changes in x away from 16.

    c. Write a mathematical statement (inequality) to describe the possible values of x.

    d. Represent the possible values of x on the number line.

3. Suppose we want the change in x away from 5 to be positive but no more than 3.

    a. List some values of x that meet this constraint.

    b. Write a mathematical statement (inequality) to describe the possible changes in x away from 5.

    c. Write a mathematical statement (inequality) to describe the possible values of x.

    d. Represent the possible values of x on the number line.

4. Suppose we want to choose a value of x that is at least 4 units away from $x = -3$.
   a. List some values of x that are at least 4 units away from $x = -3$, then represent all possible values on the number line.

   b. Write a mathematical statement (inequality) that describes values of x at least 4 units away from $x = -3$. Try to think of at least two different ways to write this.

Absolute Value

Given a real number $c > 0$,

- $|x - h| = c$ represents all values of x exactly c units away from h on the real number line.
- $|x - h| < c$ represents all values of x less than c units away from h on the real number line.
- $|x - h| > c$ represents all values of x more than c units away from h on the real number line.

---

5. Let's revisit your answers to the previous exercises. If you didn't already, use absolute value notation to rewrite the solutions to Exercises #1c, #2b, and #4b.

*6. Jamie is assembling a model airplane that flies. The instructions say that the best final weight is 28.5 ounces. However, the actual weight can vary from this by up to 0.6 ounces without affecting how well the model flies.
   a. Represent the weights that are "acceptable" on the number line and describe them using an inequality.

   b. Write a mathematical statement (inequality) that represents the acceptable variation in the model's weight
      i. using absolute value.   ii. without using absolute value.

7. a. What is the meaning of the statement "$|x - 5| < 2$"?

   b. Represent all values of x that make the statement true on the number line.

   c. Rewrite the statement in part (a) without using absolute value.

---

8. a. What is the meaning of the statement "$(w-1) \geq 5$ or $(w-1) \leq -5$"?

    b. Represent all values of w that make the statement true on the number line.

    c. Rewrite the statement in part (a) using absolute value.

*9. Use absolute value notation to represent the following.
    a. All numbers x whose distance from 6   b. All numbers r whose distance from -2 is
        is no more than 7.5 units.   more than 1 unit.

*10. For each statement, do the following.
     i. Describe the meaning of the statement.
     ii. Rewrite the statement without using absolute value.
     iii. Determine the values of the variable that make each statement true.
     iv. Represent the solutions on a number line.
    a. $|x-11| < 6$ b. $|y+7| \geq 1$

    c. $|a-3.3| > 0.8$ d. $|p+2| \leq 5$

11. a. Use the definition of absolute value to describe the solutions for each of the following statements, then state the values of x that make the statement true.

 i. $|x - 0| = 6$

 ii. $|x - 0| < 3$

 b. Based on your work in part (a), what does each of the following represent for a real positive number c?

 i. $|x| = c$

 ii. $|x| \geq c$

12. For each statement, do the following.
 i. Describe the meaning of the statement.
 ii. Write the statement without using absolute value.
 iii. Represent the solutions on the number line.

 a. $|x| = 4$

 b. $|x| \leq 5.5$

 c. $|x| > 3$

I. QUANTITIES AND CO-VARIATION OF QUANTITIES (Text: S1)

1. Consider a cup of coffee.
   a. Identify five attributes of the cup of coffee that can be measured. Which of these attributes are fixed and which are varying?
   b. Identify five attributes of the cup of coffee that cannot be measured.

2. A group of students are taking an exam. You walk into the room and identify some elements and aspects of the situation. Determine if the following statements define a quantity. If not, rewrite the statement so that the statement properly defines a quantity.
   a. the number of people taking the exam
   b. the people sitting around a table
   c. the exam
   d. the questions on the exam
   e. the teacher

3. For the following situations identify one constant quantity and two varying quantities. Define variables to represent the values of the varying quantities.
   a. A mountain climber hikes with two friends for 5 hours.
   b. The computer charges for 1.25 hours each night.
   c. The student studies for 8 hours each weekend.
   d. Jessica bikes 30 miles around a 5-mile course.

4. For the following situations identify the quantities whose values vary and the quantities whose values are constant. State possible units for measuring each of these quantities. Then define variables to represent the values of each varying quantity.
   a. A 10-inch candle burns for 2 hours.
      i. Identify at least one constant quantity and state the units of measurement.
      ii. Identify at least two varying quantities and state the units of measurement.
      iii. Define variables to represent the values of the varying quantities you defined in part (ii).
   b. A girl runs around a ¼-mile track.
      i. Identify at least one constant quantity and state the units of measurement.
      ii. Identify at least two varying quantities and state the units of measurement.
      iii. Define variables to represent the values of the varying quantities you defined in part (ii).
   c. A scuba diver descends from the surface of the water to a depth of 60 feet.
      i. Identify at least one constant quantity and state the units of measurement.
      ii. Identify at least two varying quantities and state the units of measurement.
      iii. Define variables to represent the values of the varying quantities you defined in part (ii).

5. The number of questions on an exam varies with the number of minutes to take the exam. For each question there are 5 minutes allotted. Write a formula that relates the number of minutes to take the exam to the number of questions on the exam. (*Be sure to define variables to represent the values of the varying quantities.*)

6. There are twelve times as many football players on a football team as there are coaches. Write a formula that relates the number of coaches to the number of players on the team. (*Be sure to define variables to represent the values of these varying quantities*).

---

7. This graph relates the total value of world exports
   (internationally traded goods) in billions of dollars to
   the number of years since 1950.
   a. Define variables to represent the values of the
      varying quantities in this situation.
   b. Interpret the meaning of the point (17.5, 1000).
   c. As the number of years since 1950 increases from
      10 to 20 years what is the change in the number of
      years since 1950? Represent this change on the
      graph.
   d. As the number of years since 1950 increases from
      10 to 20 years what is the corresponding change in
      the total value of world exports (in billions of
      dollars)? Explain how you determined this value
      and represent this change on the graph.

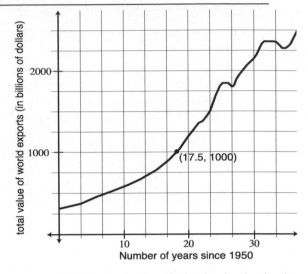

8. A ball is dropped off of the roof of a building. This graph
   relates the height of the ball above the ground (in feet) to the
   number of seconds that have elapsed since the ball was
   dropped.
   a. Define variables to represent the values of the varying
      quantities in this situation.
   b. Interpret the meaning of the point (0, 367).
   c. As the number of seconds since the ball was dropped
      increases from 10 to 30 seconds, how does the height of
      the ball above the ground change?

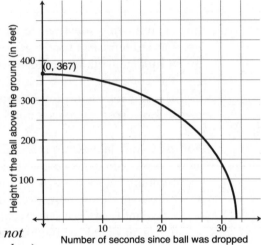

9. Write an equation for each of the relationships that are
   described below. Then solve the equation for the unknown. (*Do not
   forget to start by defining a variable to represent the unknown value.*)
   a. 17.5 is equal to 2 times some number. What is the number?
   b. The sum of 3 times some number and 12 is 42. What is the number?
   c. ¼ of some number is 4.3. What is the number?
   d. Some number is 4 times as large as 9.8. What is the number?
   e. Some number is equal to 1/3 of the sum of 88.2, 93.5, and 64. What is the number?
   f. The change from some number to 12.5 is –5. What is the number?
   g. Measuring some number in units of 12 is 3.5. What is the number?

10. Write an equation for each of the relationships that are described below. Then solve the equation for
    the unknown. (*Do not forget to start by defining a variable to represent the unknown value.*)
    a. 45 is some multiple of 15. Determine the value of the multiple.
    b. $1,200 is 1.5 times as large as some amount of money. What is the amount of money?
    c. 200% of some number is 38.2. What is the number?
    d. 5 is the result of 12 more than ¾ of some number. What is the number?
    e. Suppose 10 is the number that is 5 times as large as the value that is 4 less than the value of *x*.
       What is the value of *x*?
    f. The ratio of the change from 2 to 8 and the change from 5 to 7. What is the ratio?
    g. 7 is 3.5 times as large as some number. What is the number?

11. Evaluate the following expressions:
    a. $-2 + 5 - 12$
    b. $6 - (-4) + 1$
    c. $2(-3)(-1) - (-6)$
    d. $\dfrac{-5 + (9 - 5(4)) + 3}{5(-2) - 1}$

12. Let $y = 3.5x - 7$.
    a. Find the value of y when x is zero.
    b. What value(s) of x give a y value of 11?
    c. What value(s) of y correspond to an x value of 3?

13. Solve each of the equations for the specified variable.
    a. Given $y = 17x - 6$, solve for x when $y = 45$.
    b. Given $y = \frac{6x+5}{3}$, solve for x when $y = 2$.
    c. Given $z = \frac{12x-4}{3}$, solve for x when $z = 10$.
    d. Given $y = \frac{3.2x(6-0.5x)}{x}$, solve for x when $y = 1$.
    e. Given $y = \frac{-6a+7-2a}{2} - 3 + a$ solve for a when $y = 2$.

14. Simplify the following
    a. $\dfrac{3x^3 + 6x}{x}$
    b. $\sqrt{52x^8 y^3}$
    c. $\dfrac{2x^2 + x - 6}{x + 2}$

15. Simplify the following
    a. $\dfrac{4(x+3) + 6x - 12}{x}$
    b. $\sqrt{70x^9 y^{153}}$
    c. $\dfrac{6x^2 - 5x - 21}{(2x+3)(3x-7)}$

II. CHANGES IN QUANTITIES AND CONSTANT RATE OF CHANGE (Text: S1, 3)

16. Matthew started a regular exercise routine and his weight changed from 175 pounds to 153 pounds. What was the change in Matthew's weight?

17. After driving from Tucson to Phoenix the number of miles on your car's odometer went from 312 to 428. What was the change in the number of miles on your car's odometer?

18. After spending an hour processing emails the number of unread emails in your inbox went from 23 to 5. What is the change in the number of unread emails?

19. Let x and z represent the values of two different quantities.
    a. If the value of x decreases from $x = 2$ to $x = -5$, what is the change in the value of x?
    b. If the value of x decreases from $x = 212$ to $x = 32$, what is the change in the value of x?
    c. If the value of z decreases from $z = 2.145$ to $z = 1.234$, what is the change in the value of z?

20. Let a and b represent the values of two different quantities.
    a. If the value of a decreases from $a = 3$ to $a = -4$, what is the change in the value of a?
    b. If the value of a increases from $a = -31$ to $a = 12.2$, what is the change in the value of a?
    c. If the value of b decreases from $b = -1.15$ to $b = -4.21$, what is the change in the value of b?

21. Fill in the following tables showing the appropriate changes in the value of the variable.

s	Δs
−12.43	
0.73	
−7.3	

y	Δy
2.85	
14.3	
−1.05	

p	Δp
3.834	
−2.3	
0	

22. This morning Tom went for a run. Let d represent the number of miles that Tom has run. What is the difference in meaning between $d = 11$ and $\Delta d = 11$?

23. Use the graph to answer the following questions;
   a. Determine Δx and Δy from the point on the left to the point on the right. Illustrate the values of Δx and Δy on the graph.
   b. Determine Δx and Δy from the point on the right to the point on the left. Illustrate the values of Δx and Δy on the graph.

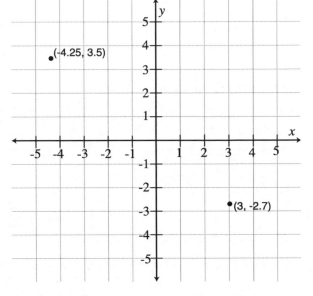

24. The number of calories Monica burns while running increases by 105 calories for every mile that Monica runs up to 15 miles. Let n represent the number of miles that Monica has run.
   a. Suppose the value of n increases $n = 3$ to $n = 7.5$.
       i.   What is the change in the value of n?
       ii. What does this change represent in the context of this situation?
       iii. What is the corresponding change in the number of calories Monica burns while running?
   b. How many calories will Monica burn for *any* 4.5-mile change in the number of miles Monica has run during a 15 mile run?
   c. For the following changes in the number of miles that Monica has run determine the corresponding change in the number of calories that Monica has burn.
       i. $\Delta n = 3.5$
       ii. $\Delta n = 0.41$
       iii. $\Delta n \doteq 13.7$
       iv. $\Delta n = k$ for some constant $k \leq 15$ miles

25. A bucket full of water has a leak. The bucket loses 71 mL of water every 5 minutes. Let t represent the number of minutes since the bucket started leaking
   a. As the number of minutes since the bucket started leaking increases from 13 minutes to 29 minutes, what is the corresponding change in the volume of water in the bucket?
   b. By how much will the volume of water in the bucket change when the number of minutes since the bucket started leaking changes by 1 minute? Explain how you determined your answer.
   c. As the volume of water in the bucket decreases from 60 mL to 23 mL, what is the corresponding change in the number of minutes since the bucket started leaking?
   d. By how much will the number of minutes since the bucket started leaking change when the volume of water in the bucket decreases by 1 mL?

26. A group of Kansas University students were traveling from Lawrence, KS to Denver, CO for a weekend ski trip. On the way they stopped for a late dinner, then continued on to Denver driving through the night. They left the restaurant, located 112 miles from Lawrence, at 10:00pm and arrived at Denver, 565 miles from Lawrence, at 5:45 am. For the purpose of this problem, assume the car maintained a constant speed from the time they left the restaurant to the time they arrived in Denver.
   a. Explain what it means to say the car maintained a constant speed from the time it left the restaurant to the time it arrived in Denver. *Make sure to explain the relationship it implies – do not say the car's speed does not change.*
   b. At what constant speed did the car travel between the restaurant and Denver?
   c. The driver was listening to music to keep awake. Between 2:03 am and 2:55 am the driver listened to his favorite album.
      i. What was the change in the time elapsed while he listened to the album in minutes? In hours?
      ii. How far did the car travel while the driver listened to this album?
   d. As the Kansas University students traveled between two towns they noticed that their trip odometer reading changed from 234.6 miles from Lawrence to 302.4 miles from Lawrence.
      i. What was the change in distance from Lawrence between these two towns?
      ii. How much time elapsed while the car traveled between these two towns?
   e. Sketch a graph of the relationship between the students' distance from Lawrence (in miles) and the number of hours since the students left the restaurant. What does the slope of the graph convey about this situation?

III. CONSTANT RATE OF CHANGE AND LINEAR FUNCTIONS (Text: S2, 3)

27. You are driving on the interstate with your cruise control on at a constant speed of 64 miles per hour. Use the number lines below to determine how long it will take to drive to the next rest sop that is 16 miles away (see textbook, page 20). Explain the thinking you used to determine your answer.

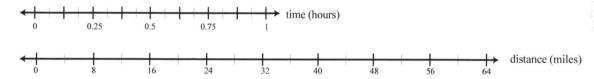

28. When an object is dropped, gravity pulls on the object and causes its speed to increase. The table below shows a certain object's speed at various moments during its fall. Does the object's speed (in feet per second) change at a constant rate with respect to the number of seconds since the object started falling? Explain your reasoning.

Number of seconds since the object started falling	The speed of the object (in feet per second)
0.15	4.83
0.4	12.88
0.52	16.744
0.98	31.556
1.26	40.572

29. The following table of values provides information about the distance of an airplane from Sky Harbor International Airport in terms of the number minutes since the plane took off. Does the distance of the airplane from Sky Harbor International Airport change at a constant rate with respect to the number of minutes since the plane took off? Explain your reasoning. If so, determine the value of the constant rate of change of the distance of the airplane from Sky Harbor International Airport in terms of the number of minutes since the plane took off.

Number of minutes since the airplane took off	Distance of the airplane from Sky Harbor International Airport (in miles)
3	16
5	32
9	64
11	92
18	170

30. Suppose that the quantities whose values are represented by x and y are related by a constant rate of change of y with respect to x.
    a. Given the information in the table determine the value of m, the constant rate of change of y with respect to x.
    b. Given the information in the graph determine the value of the constant rate of change of y with respect to x.

x	y
−3.5	7.1
−1	−0.9
2	−10.5
6	−23.3
10	−36.1

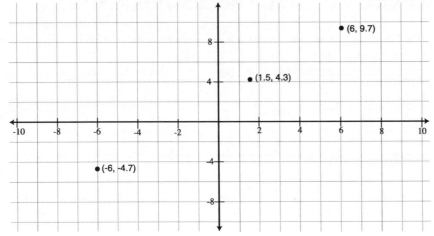

31. Given values of x and y in the tables, which table(s) contain values that could define a linear relationship between two quantities? If the two quantities whose values are represented by x and y can be related by a linear relationship, what is the constant rate of change (the value of m)?

Table 1	
x	y
−2	−14
3	1
5	7
8	16
12	28

Table 2	
x	y
1	3
2	7
4	8
5	12
9	14

Table 3	
x	y
−5	9.5
−2	5
−1	3.5
3	−2.5
10	−13

32. a. Between 2000 and 2005 the Burger Company's profit increased by $3,500 per year. In 2000 the Burger Company's profit was $52,000.
    i. Determine if the change in the number of years since 2000 is proportional to the change in the Burger Company's profit. Explain your reasoning.
    ii. Determine if the number of years since 2000 is proportional to the Burger Company's profit. Explain your reasoning.
  b. You are planning a trip to Las Vegas and need to rent a car. After contacting the different car companies, you choose to go with the company that charges a $25 rental fee and $0.05 per mile that the car is driven.
    i. Determine if the change in the number of miles driven is proportional to the change in the total cost of the rental car. Explain your reasoning.
    ii. Determine if the number of miles driven is proportional the total cost of the rental car. Explain your reasoning.
  c. When baking chocolate chip cookies, you need 3 cups of flour per cup of sugar.
    i. Determine if the change in the number cups of flour is proportional to the change in the number of cups of sugar. Explain your reasoning.
    ii. Determine if the number of cups of flour is proportional to the number of cups of sugar. Explain your reasoning.

33. Nick is considering joining a weight loss club that provides meals and support for people who want to lose weight. Based on an initial consultation with a weight loss advisor, Nick charted his potential weight loss based on the advisor's estimates of his expected weekly weight loss.

Number of weeks since joining	Nick's projected weight (pounds)
3	277
6	266.5
8	259.5
12	245.5
13	242

  a. Are the quantities Nick's projected weight (in pounds) and time since joining (in weeks) proportional? Justify your answer.
  b. Complete the following table showing the relative changes in the quantities change in time since joining (in weeks) and change in Nick's projected weight (in pounds).

Change in the number of weeks since joining	Number of weeks since joining	Nick's projected weight (in pounds)	Change in Nick's projected weight (pounds)
	3	277	
	6	266.5	
	8	259.5	
	12	245.5	
	13	242	

  c. Do the quantities change in Nick's projected weight (in pounds) and change in time since joining (in weeks) appear to be proportional? Justify your answer.
  d. Construct a graph showing the relationship between the quantities Nick's projected weight (in pounds) and time since joining (in weeks).
  e. What does the slope of the graph represent in this context?

34. For the tables below,
    a. Determine if the quantities are proportional.
    b. Determine if the changes in the quantities are proportional.
    c. Determine if the relationship is linear. If the table represents a linear relationship, write a formula to represent how the quantities change together

Table 1	
x	*y*
2	1.42
−1.2	−0.852
5.1	3.621

Table 2	
a	*b*
0.7	1.3
−2	5.62
5.4	−2.885

Table 3	
r	*s*
2.4	0.76
0.3	3.595
−1.5	6.025

IV. CONSTANT RATE OF CHANGE & LINEARITY (Text: S3)

For Exercises #35-36: Let r represent the possible values of one quantity and let p represent the possible values of another quantity.

35. Suppose r changes at a constant rate of 2 with respect to p.
    a. What does this mean for any change in p?
    b. If p changes by 6, how much does r change?
    c. If p changes by −3.1, how much does r change?

36. Suppose r changes at a constant rate of −1.3 with respect to p.
    a. What does this mean for any change in p?
    b. If p changes by 2, how much does r change?
    c. If p changes by −6.2, how much does r change?

37. Suppose the constant rate of change of y with respect to x is 0.17 and we know $y = 12.25$ when $x = 7.35$.
    a. What is the value of y when $x = 11.1$?
    b. What is the value of y when $x = −5.6$?
    c. What is the value of x when $y = 2.5$?
    d. What is the change in the value of y when the change in the value of x is 4.75?

38. Suppose the constant rate of change of y with respect to x is −11.1 and we know $y = −2.6$ when $x = −0.85$.
    a. What is the value of y when $x = 3.4$?
    b. What is the value of y when $x = −12.7$?
    c. What is the value of x when $y = 4.5$?
    d. What is the change in the value of y when the change in the value of x is −6.7?

39. Rope is wound around a large spool. The more rope wound around the spool, the greater the combined weight of the spool and rope. The graph shows that when 5 feet of rope is wound around the spool, the total weight of the spool and rope is 3.95 pounds. Note that the rope weighs 0.27 pounds per foot.

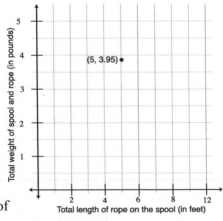

    a. If the number of feet of rope on the spool increases from the given point to 8.4 feet, what is the change in the number of feet of rope on the spool? Represent this change on the graph.
    b. By how much does the weight of the spool and rope change for the change in the amount of rope on the spool you found in part (a)? Represent this change on the graph.
    c. What is the total weight of the spool and rope when there are 8.4 feet of rope on the spool? Explain how you determined this.
    d. What is the weight of the spool without any rope? Explain how you determined this value and represent your reasoning on the graph.

40. When a bathtub made of cast iron and porcelain contains 60 gallons of water the total weight of the tub and water is approximately 875.7 pounds. You pull the plug and the water begins to drain. (*Note that water weighs 8.345 pounds per gallon*).
    a. Describe the quantities in this situation. Which of these quantities are constant and which are changing?
    b. Suppose that some water has drained from the tub and 47 gallons of water remain in the tub.
       i.   What was the change in the number of gallons of water (recall the situation begins with 60 gallons of water in the tub)?
       ii. What is the corresponding change in the total weight of the tub and water?
       iii. What is the weight of the tub and water when there are 47 gallons of water in the tub?
    c. Complete the following table of values.

Number of gallons of water remaining in the tub	Total weight of the tub and water (in pounds)
59	
40	
30	
20	
10	
8.5	

    d. Suppose you and a friend can each lift about 150 pounds. Once empty, could you and your friend pick up and carry the bathtub out of the bathroom? Explain your reasoning.

41. John inserts a partially used battery into a portable electric fan. The percent of the battery's total charge changes at a constant rate of –3.1% per minute since the fan was totally charged.
    a. Represent "an increase of 10 minutes in the amount of time the fan is used" from the given reference point on the graph.
    b. By how much will the percent of the total possible battery charge change when the fan is used for 10 minutes? Represent this on the graph.
    c. What is the percent of the battery's total possible charge when the fan has been used for 17 minutes? Explain how you determined this value.
    d. What is the vertical intercept of the graph of the function? Explain how you determined this value and represent your approach on the graph above. What does the vertical intercept represents in the context of this situation.

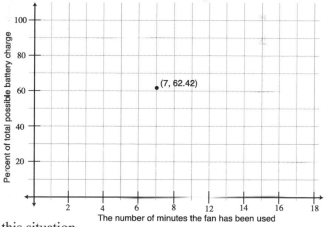

42. Consider the graph and assume that the values of x and y change together at a constant rate of change.
    a. Determine the constant rate of change of y with respect to x.
    b. Using the point $(-5.2, 1.30)$ as a reference point, what is the change in x from this point from to $x = 2$. Represent this change using the given axes.
    c. What is the change in y that corresponds with the change in x found in part (b)?
    d. What is the value of y when $x = 2$?
    e. What is the vertical intercept of the function? Explain how you can find this value using the meaning of constant rate of change and the reference point $(-5.2, 1.30)$.

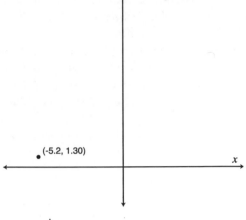

43. Consider the graph and assume that the values of x and y change together at a constant rate of change.
    a. Determine the constant rate of change of y with respect to x.
    b. Using the point $(1.8, 4.74)$ as a reference point, what is the change in x from this point from to $x = 3.6$. Represent this change using the given axes.
    c. What is the change in y that corresponds with the change in x found in part (b)?
    d. What is the value of y when $x = 3.6$?
    e. What is the vertical intercept of the function? Explain how you can find this value using the meaning of constant rate of change and the reference point $(1.8, 4.74)$

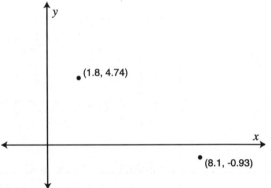

44. Two quantities (A and B) co-vary such that Quantity A changes at a constant rate with respect to Quantity B. Suppose the change in the value of Quantity A is always -0.82 times as large as the change in the value of Quantity B. Suppose also that the value of Quantity A is -1.4 when the value of Quantity B is -2.5.
    a. What is the value of Quantity A when the value of Quantity B is 3?
    b. What is the value of Quantity A when the value of Quantity B is -7.7?
    c. What is the value of Quantity B when the value of Quantity A is -6?

45. Given the values of x and y in the tables that follow, which table(s) contain values that could define a linear relationship between the two quantities? Explain your reasoning. For the table(s) that could represent a linear relationship, write a formula to define how the quantities change together.

Table 1	
x	y
1.25	3
3	6.5
5.5	11.5

Table 2	
x	y
3.25	10
3	9
5	4

Table 3	
x	y
3	10
6	20
7	70/3

46. Given the values of x and y in the tables that follow, which table(s) contain values that could define a linear relationship between the two quantities? Explain your reasoning. For the table(s) that could represent a linear relationship, write a formula to define how the quantities change together.

Table 1	
x	y
2	10.4
5	30.65
10	64.5

Table 2	
x	y
1.3	12.015
–4	14.4
9.7	8.235

Table 3	
x	y
–2	13.6
1	2.2
4.3	–10.34

47. Find a formula for each of the linear functions whose graphs are described below.
    a. The graph of the function that passes through the point $(2, -18.4)$ and the change in y is always -1.34 times as large as the change in x.
    b. The graph of the function with a constant rate of change of y with respect to x is $\frac{7}{9}$ and passes through the point $\left(\frac{-3}{5}, \frac{2}{11}\right)$.
    c. The graph of the function passing through the points $(-2, 14)$ and $(-12, -7.6)$.
    d. The graph of the function passing through the points $\left(\frac{11}{9}, \frac{15}{7}\right)$ and $\left(\frac{-12}{4}, \frac{15}{7}\right)$.

48. The graph of a certain linear function passes through the points $(2,9)$ and $(7, -11)$.
    a. What is the constant rate of change of y with respect to x (slope) for the function?
    b. From the point $(2, 9)$ how much must x change to reach a value of $x = 0$?
    c. What is the corresponding change in the value of y for the change in x you found in part (b)?
    d. What is the value of y when $x = 0$?
    e. Write a formula to calculate the value of y for any value of x.

49. The graph of a certain linear function passes through the points $(-7, -15)$ and $(5, -7)$.
    a. What is the constant rate of change of y with respect to x (slope) for the function?
    b. From the point $(-7, -15)$ how much must x change to reach a value of $x = 0$?
    c. What is the corresponding change in the value of y for the change in x you found in part (b)?
    d. What is the value of y when $x = 0$?
    e. Write a formula to calculate the value of y for any value of x.

50. Consider the formula $y = -12.13x + 7.14$. Suppose we want to find the value of y when $x = -1.15$. Explain how the formula determines the value of y using the meaning of constant rate of change. (You may sketch a graph or diagram if it helps you explain.)

51. The formula $a = 10 - 1.5t$ defines the remaining height (in inches) of a burning candle, a, in terms of the number of hours that the candle has been burning, t.
    a. What does 10 represent in the context of this situation?
    b. What does -1.5 represent in the context of this situation?
    c. What does $-1.5t$ represent in the context of this situation?
    d. Explain what the point $(t, a) = (2, 7)$ conveys in the context of this situation.
    e. What is the value of a when $t = 4.2$. Explain what this value represents in the context of this situation.

52. Determine a formula that defines the linear functions whose graphs are described below.
    a. The graph of the function conveys that the constant rate of change of y with respect to x is $\frac{2}{3}$ and the vertical intercept is 5.
    b. The graph of the function conveys that the constant rate of change of y with respect to x is 5 and the graph crosses the vertical axis at $(0, -2)$
    c. The graph of the function conveys that the constant rate of change of y with respect to x is $-\frac{6}{7}$ and the vertical intercept of $-\frac{1}{10}$.

53. Write the formula that defines the linear relationship given in each of the following graphs.
    a.   b.

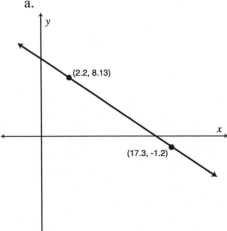   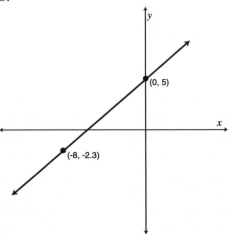

54. Use the graph to answer the following questions.
    a. The graph conveys that the constant rate of change of y with respect to x is -4.1. What is the value of y when $x = 5.15$? Explain how to find this value using the meaning of constant rate of change. Draw a diagram to help represent your reasoning.
    b. What is the value of y when $x = -3.81$? Explain how to find use the idea of constant rate of change to find the value of y.

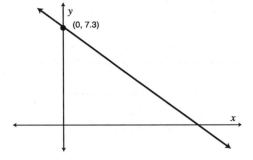

55. A tortoise and hare are competing in a race around a 1600-meter track. The arrogant hare decides to let the tortoise have a 630-meter head start. When the start gun is fired, the hare begins running at a constant speed of 8.5 meters per second and the tortoise begins crawling at constant speed of 6 meters per second.
    a. What quantities are changing in this situation? What quantities are not changing? (Be sure to include the units of each quantity.)
    b. Define a formula to determine the distance of the tortoise from the starting line in terms of the amount of time since the start gun was fired.
    c. Define a formula to determine the distance of the hare from the starting line in terms of the amount of time since the start gun was fired.
    d. The tortoise traveled 170 meters as he moved from his starting position to a curve on the track. If possible, find the following:
        i.   The amount of time that it took the tortoise to travel the 170-meter distance.
        ii. The amount of time that it took the tortoise to travel the next 80 meters on the track.

(Exercise continues on the next page.)

   e.   Now consider how the distance between the tortoise and the hare changes throughout the race.
       i.   Explain how the distance between the tortoise and hare changes as the number of seconds since the start gun was fired increases.
      ii.   Define a formula that relates the distance between the tortoise and the hare with the number of seconds since the start gun was fired.
    iii.   Is the relationship you defined in part (ii) linear? Explain.
    iv.   Who finishes the race first, the tortoise or the hare? Explain.

56. Lisa and Sarah decided to meet at a park bench near both of their homes. Lisa lives 1850 feet due west of the park bench and Sarah lives 1430 feet due east of the park bench. Sarah left her house at 7:00 pm and traveled at a constant speed of 315 feet per minute towards the bench. Lisa left her house at the same time traveling a constant speed of 325 feet per minute towards the bench.
   a.   Illustrate this situation with a drawing, labeling the constant and varying quantities.
   b.   Define a formula to relate Lisa's distance (in feet) from the park bench in terms of the number of minutes that have passed since 7:00 pm. (Define relevant variables.)
   c.   Define a formula to relate Sarah's distance (in feet) from the park bench in terms of the number of minutes that have passed since 7:00 pm. (Define relevant variables.)
   d.   Who will reach the park bench first? Explain your reasoning.
   e.   Now we will consider how the distance between Lisa and Sarah changes as the number of minutes since 7:00 pm increases.
       i.   Explain how the distance (in feet) between Lisa and Sarah changes as the amount of minutes since 7:00pm increases.
      ii.   Define a formula that relates the distance between Lisa and Sarah in terms of the number of minutes since 7:00pm.

57. John and Susan leave a neighborhood restaurant after having dinner. They each walk to their respective homes. John's home is 3120 feet due north of the restaurant and Susan's home is 2018 feet due south of the restaurant. John leaves at 7:28 pm traveling at a constant speed of 334 feet per minute and Susan leaves at 7:30 pm traveling at a constant speed of 219 feet per minute.
   a.   Illustrate this situation with a drawing and define relevant variables.
   b.   Define a formula to relate John's distance from the restaurant in terms of the number of minutes that have elapsed since 7:30 pm.
   c.   Define a formula to relate Susan's distance from the restaurant in terms of the number of minutes that have passed since 7:30 pm.
   d.   Will John or Susan arrive home first? Explain your reasoning.
   e.   Now we will consider how the distance between John and Susan changes as the number of minutes that have passed since 7:30 pm increases.
       i.   Explain how the distance (in feet) between John and Susan changes as the number of minutes that have passed since 7:30 pm increases.
      ii.   Define a formula that relates the distance (in feet) between John and Susan in terms of the number of minutes that have passed since 7:30pm.

58. Lucia rented a car for $211 per week (with unlimited driving miles) to use during her spring break vacation. She must also pay for gasoline which costs $3.65 per gallon. The gas mileage of the car is 30 miles per gallon (mpg) on average.
   a.   What quantities are changing in this situation? What quantities are not changing?
   b.   i.   If Lucia traveled 100 miles, how many gallons of gasoline did she use?
      ii.   If Lucia traveled 200 miles, how many gallons of gasoline did she use?
     iii.   If Lucia traveled 500 miles, how many gallons of gasoline did she use?

(Exercise continues on the next page.)

   c.   Define a formula to determine the number of gallons of gasoline n used in terms of the number of miles driven x.

   d.   Define a formula to determine the cost of gasoline c in terms of the number of gallons of gasoline used.

   e.   Define a formula to determine the total cost T of driving the rental car x miles, including the one-week rental cost of $211.

   f.   How much does the total rental cost increase for each 100 miles the car is driven during the rental period. Explain your reasoning.

   g.   How much does the total rental cost increase for each mile driven? Explain the thinking you used to arrive at your answer.

59. Sketch a graph of the following relationships.
   a. $3.25 = y$
   b. $6 = x$
   c. $y = 2x + 4.3$
   d. $y = -\frac{1}{2}x + 3$

60. On the same axes sketch a graph of $y = -4x - 2$ and $y = \frac{1}{4}x + 3$. What do you notice about the graphs of two linear relationships?

61. Simplify the following expressions.
   a. $2x + 7 - 3x - 2$
   b. $\frac{3}{7}x - (-1 + \frac{2}{3}x)$
   c. $3x - (-7x) + 4 - 2.2$
   d. $2(x - 4) + 3x - (\frac{9}{8}x - 7)$

V. Exploring Average Speed (Test: S4)

62. This graph represents the distance-time relationship for Kevin and Carrie as they cycled on a road from mile marker 225 to mile marker 230.

   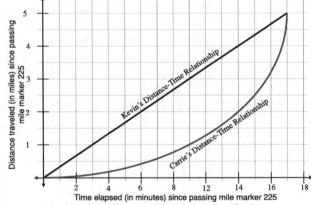

   a.   How does the distance traveled and time elapsed compare for Carrie and Kevin as they traveled from mile marker 225 to mile marker 230?
   b.   How do Carrie's and Kevin's speeds compare as they travel from mile marker 225 to mile marker 230?
   c.   How do Carrie's and Kevin's average speeds compare over the time interval as they traveled from mile marker 225 to mile marker 230?
   d.   Do Carrie and Kevin collide on the course 17 minutes after the passed mile marker 225?

63. When running a marathon you heard the timer call out 12 minutes as you passed mile-marker 2.
   a.   What quantities could you measure to determine your speed as you ran the race? Define variables to represent the quantities' values and state the units you will use to measure the value of each of these quantities.
   b.   As you passed mile-marker 5 you heard the timer call out 33 minutes. What was your average speed from mile 2 to mile 5?
   c.   Assume that you continued running at the same constant speed as computed in (b) above. How much distance did you cover as your time spent running increased from 35 minutes after the start of the race to 40 minutes after the start of the race?
   d.   If you passed mile marker 5 at 33 minutes, what average speed do you need to run for the remainder of the race to meet your goal to complete the 26.2-mile marathon in 175 minutes?
   e.   What is the meaning of average speed in this context?

64. When running a road race you heard the timer call out 8 minutes as you passed the first mile-marker in the race.
    a. What quantities will you measure to determine your speed as you travel? Define variables to represent the quantities' values and state the units you will use to measure the value of each of these quantities.
    b. As you passed mile-marker 6 you heard the timer call out 52 minutes. What was your average speed from mile-marker 1 to 6?
    c. After mile-marker 6 you slowed down and ran at a constant rate of 10 minutes per mile between mile-marker 6 and mile-marker 10, how many minutes did it take you to travel from mile marker 6 to 9?

65. Marcos traveled in his car from Phoenix to Flagstaff, a distance of 155 miles.
    a. Determine the amount of time required for Marcos to travel from Phoenix to Flagstaff if his average speed for the trip was 68 miles per hour.
    b. Construct a possible distance-time graph of Marcos's trip from Phoenix to Flagstaff. Be sure to label your axes.
    c. On the same axes, construct a graph that represents the distance-time graph that represents another car traveling at a constant speed for the entire trip.

66. On a trip from Tucson to Phoenix via Interstate 10, you used your cruise control to travel at a constant speed for the entire trip. Since your speedometer was broken, you decided to use your watch and the mile markers to determine your speed. At mile marker 219 you noticed that the time on your digital watch just advanced to 9:22 am. At mile marker 197 your digital watch advanced to 9:46 am.
    a. Compute the constant speed at which you traveled over the time period from 9:22 am to 9:46 am.
    b. As you were passing mile marker 219 you also passed a truck. The same truck sped by you exactly at mile marker 197.
       i. Construct a distance-time graph of your car. On the same graph, construct one possible distance-time graph for the truck. Be sure to label the axes.
       ii. Compare the speed of the truck to the speed of the car between 9:22 am and 9:45 am.
       iii. Compare the distance that your car traveled over this part of the trip with the distance that the truck traveled over this same part of the trip. Compare the time that it took the truck to travel this distance with the time that it took your car to travel this distance. What do you notice?
       iv. Why are the average speed of the car and the average speed of the truck the same?
       v. Phoenix is another 53 miles past mile marker 197. Assuming you continued at the constant speed, at what time should you arrive in Phoenix?

67. The distance d (measured in a number of feet) between Silvia and her house is modeled by the formula, $d = t^2 + 3t + 1$ where t represents the number of seconds since Silvia started walking.
    a. Find Silvia's average speed for the time period from $t = 2$ to $t = 7$ seconds.
    b. What was Silvia's change in distance as the time since Silvia started walking increased from 2 to 7 seconds?
    c. i. Construct a graph that gives Silvia's distance from her house (in feet) in terms of the number of seconds since she started walking. Be sure to label your axes.
       ii. Illustrate (with a line segment on the graph you constructed in part (i)) Silvia's change in distance during the time period from $t = 4$ and $t = 5$ seconds since she started walking.
       iii. Illustrate (with a line segment on the graph you constructed in part (i)) Silvia's change in distance during the time period from $t = 5$ and $t = 6$ seconds since she started walking.

68. Bob's distance, d, north of Mrs. Bess's restaurant (in feet) is given by the formula $d = 2t^2 - 7$ where t represents the number of seconds since Bob began driving.
    a. Determine the value of d when $t = 1$. What does a negative value for d represent in the context of this problem?
    b. Find the average speed of Bob's car for the time period from $t = 3$ to $t = 5$.
    c. As the number of seconds since Bob began driving increased from 1.5 to 2 seconds, by how much did Bob's distance north of Mrs. Bess's restaurant change?
    d. As the number of seconds since Bob began driving increased from 2 to 2.5 seconds, by how much did Bob's distance north of Mrs. Bess's restaurant change?
    e. i.   How much time did it take Bob to travel from 20 to 30 feet north of the restaurant?
        ii. How much time did it take Bob to travel from 30 to 40 feet north of the restaurant?
        iii. How much time did it take Bob to travel from 40 to 50 feet north of the restaurant?

69. The graph represents the speeds of two cars (car A and car B) in terms of the elapsed time in seconds since being at a rest stop. Car A is traveling at a constant speed of 65 miles per hour. As car A passes the rest stop car B pulls out beside car A and they both continue traveling down the highway.
    a. Which graph represents car A's speed and which graph represents car B's speed? Explain.
    b. Which car is further down the road 20 seconds after being at the rest stop? Explain.
    c. Explain the meaning of the intersection point.
    d. What is the relationship between the positions of car A and car B 25 seconds after being at the rest stop?

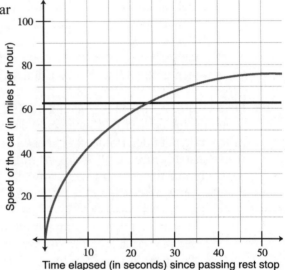

For Exercises #70-79, let d be the distance of a car (in feet) from mile marker 420 on a country road and let t be the time elapsed (in seconds) since the car passed mile marker 420. The formulas below represent various ways these quantities might be related. For each of the following:
    i.   Determine the average speed of the car using the given formula and the specified time interval.
    ii. Explain the meaning of average speed for the given situation.

70. $d = t^2$ from $t = 5$ to $t = 30$.

71. $d = -3(-19t - 1)$ from $t = 3$ to $t = 9$.

72. $d = 5(12t + 1) + 3t$ from $t = 0.5$ to $t = 3.75$.

73. $d = \frac{10t(t+5) - 14}{2}$ from $t = 0$ to $t = 5$.

74. $d = \frac{1}{3}\left(9t^2 + 155t - (11t - 6)\right)$ from $t = 2$ to $t = 4$.

75. $d = (2t + 7)(3t - 2)$ from $t = 2$ to $t = 2.75$.

76. $d = \left(\frac{1}{3}t + 60\right)\left(t + \frac{1}{2}\right)$ from $t = 1$ to $t = 4$.
    from $t = 30$ to $t = 35$.

77. $d = (t + 6)(t + 3) + 7t - 20 + 11t - \frac{7}{8}t^2$
    from $t = 30$ to $t = 35$.

78. $d = \frac{1}{8}t\left(3t^2 + 1.5t\right) + 16t - 3$ from $t = 2$ to $t = 4$.

79. $d = \dfrac{t\left(\frac{1}{2}t^2 + 15\right) + 3t\left(\frac{1}{10}t^2 + \frac{4}{3t}\right)}{5}$
    from $t = 5$ to $t = 9$.

VI. THE DISTANCE FORMULA (Text: S5)

80. Use the distance formula to find the distance between the points on the graph.

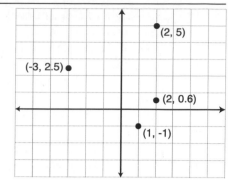

81. Three corners of a rectangle are located at the points (–2, –1), (7, 3.5) and (–2, 3.5) on the coordinate axes. What is the point where the fourth corner of the rectangle is located? Justify your answer.

82. The endpoints of one side of a rotated squared are located at (–5, –7) and (1, –1) on the coordinate axes. Find the dimensions and area of the square.

83. A circle is centered at the origin (0, 0) and has radius of length 4.
    a. For which points on the circle is the *x*-coordinate equal to 1?
    b. For which points on the circle is the *y*-coordinate equal to 3?
    c. What is the equation for the graph of this circle?

84. A circle is centered at the origin (–2, 4) and has radius of length 6.
    a. For which points on the circle is the *x*-coordinate equal to 0?
    b. For which points on the circle is the *y*-coordinate equal to 4?
    c. What is the equation for the graph of this circle?

85. The center of a circle is located at the point (1, 1). The point (–5, 2) is located on the circle.
    a. What is the radius of the circle?
    b. What is the equation for the graph of this circle?

86. The center of a circle is located at the point (4, 6). The point (12, 9) is located on the circle.
    a. What is the radius of the circle?
    b. What is the equation for the graph of this circle?

87. The center of a circle is located at the point (–5, 0). The point (–9, –3) is located on the circle.
    a. What is the radius of the circle?
    b. What is the equation for the graph of this circle?

VI. ABSOLUTE VALUE (Text: S6)

88. A contractor is digging a hole that needs to be 36" deep. He knows that his measurement is no farther than 1.5 inches different from the actual depth of the hole.
    a. Indicate on a number line the depths that the hole could be.
    b. Use algebraic symbols to represent values of *x* (hole depth values) that are less than or equal to 1.5 inches away from 36 inches (the desired depth of the hole).
    c. Use algebraic symbols to represent the difference between varying values of the depth of the hole, *x*, and 36 inches (the ideal depth of the hole), sometimes called the margin of error.
    d. Use symbols to represent values of the margin of error that are less than or equal to 1.5 inches.

89. The ideal weight of a bag of cookies is 541 grams. The actual weight may vary from the ideal by no more than 3 grams for the bag of cookies to be sold.
    a. Represent the possible weights of the bag of cookies on a number line.
    b. Use algebraic symbols to represent values of w (the weight of the bag of cookies) that are less than or equal to 3 grams away from 541 (the ideal weight of the bag of cookies).
    c. Use algebraic symbols to represent the difference between varying values of the weight of the bag of cookies, w, and 541 grams (the ideal weight of a bag of cookies), called the margin of error.
    d. Use symbols to represent values of the margin of error that are less than or equal to 3 grams.

90. a. Represent all numbers whose distance from 3 is less than 4 on a number line.
    b. What is the meaning of $-4 < x - 3 < 4$

91. a. Represent all numbers whose distance from -5 is less than 2 on a number line.
    b. What is the meaning of $-2 < x - (-5) < 2$

92. Describe the values of x that are being described by the given inequalities.
    a. $-4 < x - 2 < 4$
    b. $-1 < x - (-3.4) < 1$
    c. $x - 2.7 > 5$ and $x - 2.7 < -5$

93. Use absolute value notation to represent the following.
    a. All numbers x whose distance from 7 is less than 4.
    b. All numbers x whose distance from -2 is less than 3.4
    c. All numbers x whose distance from -5 is greater than 5.2

94. Illustrate the solutions to the given absolute value equations on a number line, then state the solutions algebraically.
    a. $|x| = 3.5$ b. $|x| < 3.5$ c. $|x| > 3.5$

95. Represent the solution set of the following absolute value inequalities by:
    i.   Describing the solution in words
    ii. Illustrating the solutions on a number line
    iii. Writing an inequality (with no absolute values).
    a. $|x - 2| < 4$ b. $|x + 7.2| < 3$ c. $|x - 14| > 5$ d. $|x + 7.3| > 2.7$

96. Solve each of the following equations for x. Check your answers and show your work.
    a. $|x| = 7$ b. $|x - 5| = 3$ c. $|4x - 7| = 15$ d. $|19.25x - 17.3| = -98.2$

    e. $|11x + 22| = 44$ f. $|2x - 2| = x + 3$ g. $|8x - 9| = |2x + 3|$ h. $|9x - 12| = |4x - 5|$

This investigation contains review and practice with important skills and procedures you may need in this module and future modules. Your instructor may assign this investigation as an introduction to the module or may ask you to complete select exercises "just in time" to help you when needed. Alternatively, you can complete these exercises on your own to help review important skills.

Evaluating and Simplifying Expressions and Solving Equations
Use this section prior to the module or with/after Investigation 1.

Given that $a = 4$, $b = 11$, $c = 5$, and $d = -4$, evaluate each of the expressions in Exercises #1-6.

1. $\dfrac{\sqrt{c-a}+b}{2}$

2. $\dfrac{(a+1)-a}{b^2-b}$

3. $\dfrac{c}{8}+2c^2-6d$

4. $\dfrac{2\sqrt{b+14}}{8-a}+5$

5. $\dfrac{d\sqrt{d+6}}{d+6}$

6. $\frac{4}{9}\sqrt{2c^2+4a}-5$

In Exercises #7-10, use the given information to write an equation and then solve it.

7. solve $y = 2x - 19$ for x if $y = 5$

8. solve $y = \sqrt[3]{x}$ for x if $y = -2$

9. solve $n = -\frac{1}{3}(p-11)+19$ for p if $n = 7$

10. solve $\dfrac{w+2}{w+6} = q$ for w if $q = -1$

In Exercises #11-14, solve each equation.

11. $\frac{2}{3}x+\frac{1}{2}=\frac{17}{2}$

12. $4=\frac{1}{5}(3n-7)$

13. $x+3=\dfrac{x-3}{4}$

14. $\dfrac{3}{4}x+\dfrac{5}{6}=5x-\dfrac{125}{3}$

In Exercises #15-24, write an equation based on each description and then solve the equation. *Be sure to define a variable to represent the value of the unknown number before writing each equation.*

15. 17.5 is equal to 2 times some number. What is the number?

16. The sum of 3 times some number and 12 is 42. What is the number?

17. one-fourth of some number is 4.3. What is the number?

18. The difference between some number and 14.3 is 2.1. What is the number?

19. Some number is 4 times as large as 9.8. What is the number?

20. Some number is equal to one-third of the sum of 88.2, 93.5, and 64. What is the number?

21. 45 is some multiple of 15. Set up an equation and determine the value of the multiple.

22. $1,200 is 1.5 times as large as some number. What is the number?

23. 200% of some number is 38.2. What is the number?

24. Three-fourths of some number increased by 12 is equal to 5 times the number. What is the number?

Evaluating and Simplifying Expressions and Solving Equations (with Function Notation)
Use this section with/after Investigation 2.

25. Given that $f(x) = \dfrac{x}{4} + x^2$, evaluate each of the following.

    a. $f(12)$ b. $f(10)$ c. $f(-2)$ d. $2 \cdot f(3)$

26. Given that $g(x) = \dfrac{3x}{x-7} + 2$, evaluate each of the following.

    a. $g(8)$ b. $g(10)$ c. $g(-1.5)$ d. $-\dfrac{g(4)}{4}$

27. Given that $h(x) = \sqrt{2x} - x$, evaluate each of the following.

    a. $h(50)$ b. $h(14)$ c. $h(18) - h(12.5)$ d. $\dfrac{h(8) - h(2)}{8 - 2}$

28. Given that $f(x) = \dfrac{x}{4} + x^2$, write the expression that represents each of the following.

    a. $f(2r)$ b. $f(p+r)$ c. $f(3x)$

29. Given that $g(x) = \dfrac{3x}{x-7} + 2$, write the expression that represents each of the following.

    a. $g(n-2)$ b. $g(4x)$ c. $g(a-b)$

In Exercises #30-33, use the given information to write an equation and then solve it.

30. if $f(x) = \sqrt{x}$, find the value of x such that $f(x) = 4$

31. if $g(x) = \dfrac{2x - 6}{5}$, find the value of x such that $g(x) = 3$

32. if $h(x) = 3(7 - x) + 4$, find the value of x such that $h(x) = 31$

33. if $j(x) = \dfrac{1}{x + 2}$, find the value of x such that $j(x) = -0.5$

Modeling and Additional Content
Use this section prior to the module or with/after any of the first three investigations.

34. a. Indicate on the figure and describe the attributes of this circle that represent the following quantities.

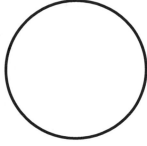

        r = radius of the circle measured in feet
        d = diameter of the circle measured in feet
        C = circumference of the circle measured in feet
        A = area of the circle measured in square feet

    b. Write the formula to express a circle's diameter in terms of its radius. Use the formula to determine the diameter of a circle that has a radius of 4.721 feet.

    c. Write the formula to express a circle's circumference in terms of its diameter. Use the formula to determine the circumference of a circle that has a diameter of 6.48 feet.

    d. Write the formula to express a circle's area in terms of its diameter. Use the formula to determine the area of a circle that has a diameter of 4.09 feet.

    e. Write the formula to express a circle's circumference in terms of its radius. Use the formula to determine the circumference of a circle that has a radius of 3.5 feet.

    f. Write the formula to express a circle's area in terms of its diameter. Use the formula to determine the area of a circle that has a diameter of 3.5 feet.

    g. Define a formula to express the circumference of a circle in terms of its area. Use the formula to determine the circumference of a circle that has an area of 42.7 square feet.

35. The distance between some number(s) and 0 on the number line is 4.
    a. What are the numbers?

    b. Illustrate on the number line all numbers that are 4 units away from 0.

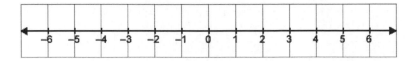

    c. Solve the equation $|x| = 4$ for x (*what value(s) of x make this equation true*)?

    d. Describe how the solutions found in part (c) are represented on the number line.

36. a. Given that $|x| = 5$, determine the value(s) of x that makes this equation true.

    b. Given that $|x - 2| = 7$, determine the value(s) of x that makes this equation true.

    c. Given that $|x + 1| = 3$, determine the value(s) of x that makes this equation true.

37. a. This graph that represents how x and y are related, determine the value of x when $y = 8$. Describe your approach and say why it works.

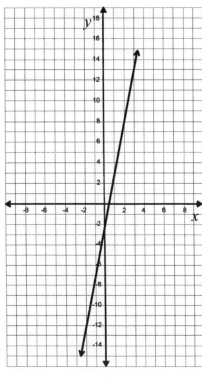

    b. Given that $y = 5x - 2$ is the equation of the graphed function, solve the equation $8 = 5x - 2$ for x algebraically. Illustrate how to use the graph to solve this equation.

    c. Use the graph to solve the equation $3 = 5x - 2$. Illustrate how you used the graph to solve this equation.

    d. Use the graph to solve the equation $0 = 5x - 2$. Illustrate how you used the graph to solve this equation.

*1. Consider what is involved in building a box (without a top) from an 8.5" by 11" sheet of paper by cutting squares from each corner and folding up the sides.

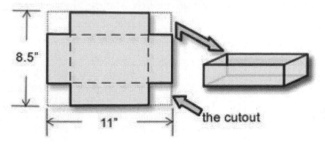

the cutout

a. To understand how the quantities in the situation are related it is important to first model the situation by doing the following.
   i. Cutting four equal-sized squares from the corners of an 8.5 by 11-inch sheet of paper.
   ii. Folding up the sides and taping them together at the edges.

b. Do the cutouts have to be square? Explain.

c. What quantities in this situation vary? What quantities in the situation are constant (do not vary)?

d. Describe how the configuration of the box changes as the length of the side of the square cutout increases. What is the largest value that makes sense for the cutout length?

e. Using the "Volume w/ Cubes" animation, describe how the volume of the box varies as the length of the side of the cutout varies from 0 to 4.25 inches.

f. Based on your response to part (e), draw a rough sketch of a graph of the volume of the box (in cubic inches) in terms of the length of the side of the square cutout (in inches). *Be sure to scale and label your axes.*

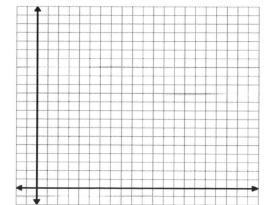

g. Use a ruler to measure the length of the side of the square cutout (in inches) of your box.
   Answer: _____

h. Determine the values of each of the following for your box.

   Box's height: _____   Length of box's base: _____

   Width of the box's base: _____   Volume of the box: _____

*2. Let x represent the varying length of the side of the square cutout in inches. Let w represent the varying width of the box's base in inches. Let l represent the varying length of the box's base in inches. Let V represent the varying volume of the box in cubic inches.

a. Complete the table of values.

x	w	l	V
0			
	6.5		
2.5			
4.25			
		2	

b. Explain how the length of the base of the box is related to the length of the side of the square cutout and how the width of the base of the box is related to the length of the side of the square cutout.

c. Construct an *expression* to represent the width of the box in terms of the length of the side of the square cutout, x.

d. Construct an *expression* to represent the length of the box in terms of the length of the side of the square cutout, x.

e. Define a *formula* to determine the volume of the box in terms of the length of the side of the square cutout.

f. Use your formula from part (e) to represent the volume of the box when x, the length of the side of the cutout, is 0.5 inches.

g. Use your formula from part (e) to represent the volume of the box when x, the length of the side of the cutout, is 3 inches.

h. Estimate the value of x, the length of the side of the square cutout, that corresponds with the maximum value for the box's volume.

3. a. View the "Graphs" animation (or use your graphing calculator) to create a graph that represents the volume of the box V (measured in cubic inches) in terms of the length of the side of the square cutout x (measured in inches).

    (When determining the window setting on your calculator, consider the possible values of x and the possible values of V.)

    Construct the graph and label two points on the graph. State what each of these points convey about the box.

    Point 1:

    Point 2:

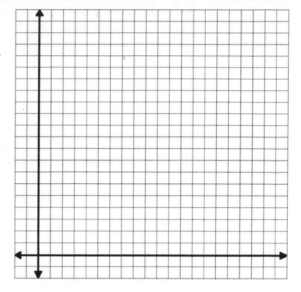

   b. Identify the point on the graph that corresponds to the dimensions of the box you created in Exercise #1.

   c. As x (the length of the side of the cutout) increases from 0.5 to 0.75 inches, how does the volume of the box change?

   d. As x (the length of the side of the cutout) increases from 1 to 3 inches, how does the volume of the box change?

   e. As x (the length of the side of the cutout) increases from 2.1 to 2.7 inches, how does the volume of the box change?

   f. Indicate on the graph i) a change of cutout length from 1 inch to 2.5 inches and ii) the corresponding change in the box's volume.

   g. Estimate the interval(s) of values for the length of the side of the cutout x for which the volume of the box decreases as x increases.

*4. a. Using your graphing calculator, solve for the length of the side of the square cutout if the volume of the box is 62.5 cubic inches.

b. Using your graphing calculator, solve for the length of the side of the square cutout if the volume of the box is 23 cubic inches.

c. Using your graphing calculator, find an approximate maximum value for the box's volume.

d. Using your graphing calculator, solve for the length of the side of the square cutout that produces the maximum volume you found in part (c).

5. An open box is constructed by cutting four equal-sized square corners from a 12-inch by 18-inch sheet of cardboard and folding up the sides. The given graph represents the volume of a box (measured in cubic inches) in relation to the length of the side of the square cutout (measured in inches).

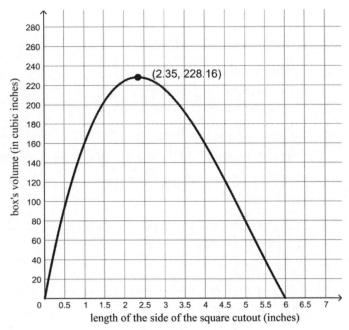

a. Determine the box's volume for each given length of the side of the square cutout. *Use the graph to check that your answer makes sense.*
   i. 2 inches

   ii. 5 inches

   iii. 3.75 inches

b. Write a formula to represent the box's volume V (in cubic inches) in terms of x, the length of the side of the square cutout (in inches).

c. Use the graph to estimate the length(s) of the side of the square cutout that produces each of the following volumes.
   i. 160 cubic inches   ii. 190 cubic inches   iii. 70 cubic inches

d. Use the graph to determine the interval(s) for the length of the side of the square cutout over which the box's volume is decreasing as the cutout length increases. Then repeat for the interval(s) over which the box's volume is increasing as the cutout length increases.

In the previous investigation we wrote formulas to describe how values of one quantity are related to values of another quantity. We say that the dependent quantity is a **function** of the independent quantity if every value of the independent quantity produces **exactly one** value of the dependent quantity. In this investigation we will introduce the uses and benefits of function notation and practice representing formulas (that assign each value of the independent quantity to exactly one value of the dependent quantity) using function notation.

It is also convention to say that values of the independent quantity are input to ("put into") the function rule, and a value produced from applying the function rule are output values of the function rule.

*1. In the previous investigation we defined the volume of a box formed by cutting squares from an 8.5" by 11" sheet of paper and folding up the sides with the formula:

$$V = x(11 - 2x)(8.5 - 2x)$$

a. Determine the value of the box's volume V when:

$x =$ 1.5 inches, $V =$ _____ cubic inches

$x =$ 3.1 inches, $V =$ _____ cubic inches

b. Discuss the following questions with your classmates.

    i. Is it possible to reference the value of V when $x = 1.5$ without calculating its value?

    ii. Is it possible to reference the formula without writing the formula?

c. ***Function notation provides a concise way to reference the value of a dependent quantity that is associated with a particular independent quantity, without actually computing its values.***

For the volume formula for the box problem, we write $f(x) = x(11 - 2x)(8.5 - 2x)$, instead of $V = x(11 - 2x)(8.5 - 2x)$.

The symbols $f(x)$ are read "f of x". The symbol f is designated to be the function name, and $f(x)$ represents the values of the function's volume. The symbols $f(2.5)$ reference the box's volume when the square's side length, $x = 2.5$ inches.

    i. What does $f(1.5)$ represent in the context of the box problem? How do you read $f(1.5)$?

    ii. What does $f(3.1)$ represent in the context of the box problem? How do you read $f(3.1)$?

    iii. What does $f(3.1) - f(1.5)$ represent? How do you read $f(3.1) - f(1.5)$?

Formulas with Function Notation

Let *x* represent the length of the side of the square cutout (in inches) and *V* represent the box's volume (in cubic inches). Call the relationship between these quantities *f* where *x* represents input values and *V* represents output values. Then:

$$f(x) = x(11 - 2x)(8.5 - 2x) \quad \text{with} \quad f(x) = V$$

Let's clarify the meaning of this notation.

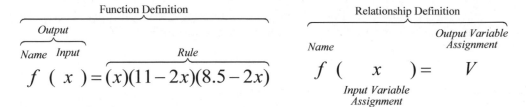

When a function is defined with a formula we produce a rule that tells us how to match each value in the domain (value of the independent quantity) to the corresponding value in the range (value of the dependent quantity).

*2. Read the above "Formulas with Function Notation" box, then answer the following questions.

    a.   Using function notation and the function name *f*, we define the box's volume, *f*(*x*), in terms of the length of the side of the square's cutout, *x*, as $f(x) = x(11 - 2x)(8.5 - 2x)$.

        i.   What is the independent (input) quantity?   ii.   What does *f*(*x*) represent?

        ii.   What does *f* represent?   iv.   What is the domain of *f*?

    b.   Illustrate each of the following on the given graph of *f*.

        i.   the point $(3.5, f(3.5))$

        ii. $f(1.5)$

        iii. the solutions to $f(x) = 40$

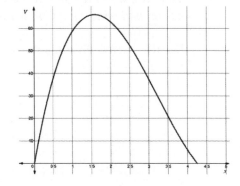

    c.   Use the function formula $f(x) = x(11 - 2x)(8.5 - 2x)$ to evaluate $f(2)$ and describe what this value represents.

    d.   Use the graph in part (b) to solve $f(x) = 60$ for *x* and describe what the solutions (values of *x*) represent. (Your answer will be approximate.)

    e.   Represent algebraically and on the given graph the change in the box's volume as the length of the side of the square cutout increases from 2 inches to 3.5 inches.

> ### Function
>
> A **function** consists of three parts:
> 1. *Domain*: The values the independent quantity may assume.
> 2. *Range*: The values the dependent quantity assume.
> 3. *Rule*: The rule assigns to each value of the independent quantity *exactly one* value of the dependent quantity.
>
> The rule of a function can be expressed using any of: i) a worded description; ii) an algebraic expression; iii) a graph; or iv) a table of values. The rule of the function defines how a function's independent and dependent quantities are related as their values change together.

Terminology: If we want to write a function f to determine the perimeter of a square P (in inches) in terms of its side length s (in inches), we write $f(s) = 4s$. (It is noteworthy that $P = f(s)$ since both P and $f(s)$ represent the varying value of the square's perimeter.)

We say that the *function f* "defines $f(s)$ in terms of s" meaning the value of the dependent quantity (the perimeter), represented by $f(s)$, is determined by (or depends on) the value of the independent quantity, s.

Some advantages of using function notation include:
- We can reference the function using its name, f. This is more concise than having to say or write out $P = 4s$.
- We can reference a function's output value that corresponds to a particular input value without having to actually compute the value. So, $f(3.2)$ represents the perimeter of a square (in inches) that has a side length of 3.2 inches.

---

*3. Read the definition and terminology overview above.
    a. What does $f(5.8)$ represent?

    b. Given that the square's perimeter cannot exceed 48 inches,
       i. what are the possible values that the square's side length s can take on?
         (*Your answer describes the **domain** of f.*)

      ii. what are the possible values that the perimeter P can take on?
         (*Your answer describes the **range** of f.*)

    c. What is the rule that determines how s and $f(s)$ are related and change together?

    d. T or F: In the function f there is exactly one value of $f(s)$ for each value of s.
       Justify your answer

*4. Billy is walking from the front door of his house to his bus stop, which is 960 feet away from his front door. As Billy walks out his front door he walks in a straight path toward his bus stop at a constant rate of 7.5 feet per second.

   a. Illustrate the situation with a diagram and define the independent variable. (Represent fixed quantities with a solid line and varying quantities with a dashed vector.)

   b. Define a function h to determine Billy's distance from his bus stop in terms of the number of seconds he has been walking.

   c. What is the independent quantity and what is the domain of h (the values the independent quantity can take on)?

   d. What is the dependent quantity and what is the range of h (the values the dependent quantity can take on)?

   e. What do each of the following represent: $h(0)$ and $h(60.25)$?

   f. Use function notation to represent each of the following quantities.
      i. Billy's distance from the bus stop after he has walked 23.6 seconds

      ii. the change in Billy's distance from the bus stop as the number of seconds since Billy left his front door increases from 35 seconds to 48 seconds

   g. If t represents the number of seconds since Billy left his front door, solve $h(t) = 150$ for t and say what your answer represents.

The Vertical Intercept of a Function

A function's *vertical intercept* is the function's output value when 0 is input to the rule of f.

We also say that the value of $f(0)$ is the *vertical intercept* of f.

It is noteworthy that if $f(0) = -25$, the vertical intercept of f is -25 and occurs at the point $(0, -25)$ on the graph of f.

---

*5 a. The function h defined by $h(x) = 39 - 6.5x$ represents the distance (in meters) that a Tortoise is ahead of a Hare in terms of the number of seconds, x, since the start of a 100 meter race. Construct the graph of h. *Be sure to label your axes.*

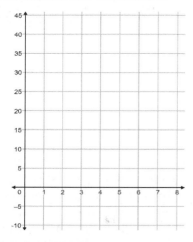

b. Determine the vertical intercept of h and describe what it represents in the context of this problem.

Horizontal Intercepts (also called Zeros or Roots) of a Function

A function f's *horizontal intercept(s)* are the input values where the graph of the function f crosses the horizontal axes.

Since a function f crosses the horizontal-axis where $f(x) = 0$, horizontal intercepts are values of x that make $f(x) = 0$. If $f(x) = 0$ when $x = a$, we say that the constant value a is a *horizontal-intercept* of f. We can also say that $x = a$ is a *root* or *zero* of the function f.

Given a function $h(x) = x(x - 7)$, we can determine the *horizontal intercepts* by determining the values of x that make $h(x) = 0$. To find the horizontal intercepts, we solve the equation $0 = x(x - 7)$. Since this equation is true when either factor is 0, the solutions are $x = 0$ and $x = 7$. The values $x = 0$ and $x = 7$ are also called *roots* or *zeros* of the function h.

*6. Continuing the same context from question 5, $h(x) = 39 - 6.5x$ represents the distance (in meters) that a Tortoise is ahead of a Hare in terms of the number of seconds, x, since the start of a 100 meter race.

a. Solve the equation $h(x) = 0$ for x. What does this solution represent in the problem context? Label this solution on the graph you created in Exercise #5, part (a).

b. What is the root of h? What point represents the horizontal intercept of the graph of h?

---

7. For each of the following functions, determine all horizontal intercepts. *Try to determine these algebraically first then verify your answers by graphing the functions on your calculator.*

    a. $h(x) = 4(x-6) + 2$ b. $b(x) = x(11 - 2x)(8.5 - 2x)$

*8. For this graphical representation of the relationship between x and y,

    a. determine if y is a function of x. Justify your answer using the definition of function—each input value of a function is assigned to exactly one output value.

    b. determine if x is a function of y. Justify your answer.

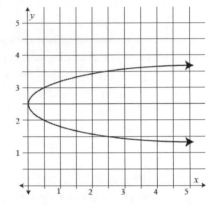

Recall that the *domain* is the set of all possible values of the independent quantity for which the function is defined. A function's domain can be restricted because:

    i) it is not possible to determine the square root of a negative number;

    ii) the denominator of a quotient cannot be 0; or

    iii) values for the independent quantity are restricted by the context (the cutout length on the box cannot exceed 4.25, for example).

9. Without using a graphing calculator determine the domain and range of the following functions.

    a. $s(x) = \dfrac{9}{x}$ b. $g(x) = \dfrac{1}{x^2 - 9}$

    c. $h(x) = x^2 + 2x - 5$ d. $k(x) = \dfrac{\sqrt{x-2}}{x-9}$

*10. Without using a graphing calculator determine the domain and range of the following functions.

    a. $p(x) = \dfrac{x}{9}$

    b. $f(x) = \sqrt{x-4}$ d.

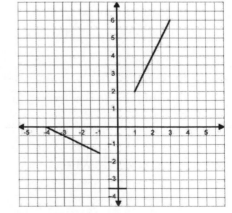

    c. $b(x) = \dfrac{x(x-5)}{x-5}$

*1. If A is the area of a square measured in square inches and s is the square's side length measured in inches, then the square's side length, s, and its area, A, are related by the formula $A = s^2$. If we call this function g, recall that using function notation we can also write $g(s) = s^2$ where $A = g(s)$.

a. Use function notation to represent (NOT CALCULATE) the areas of two different squares whose side lengths are 3.5 inches and 26.92 inches.

b. Interpret the meaning of $g(4.9) = 24.01$ in this context.

c. What does it mean to solve $g(s) = 81$ for s? (*Don't solve yet and don't explain how to solve for s, instead explain what the solution would represent.*)

d. Solve for s when $g(s) = 24.01$. (Recall that $g(s) = s^2$, where the square's area $A = g(s)$.)

e. Explain what each of the following represent.
   i. $g(5) + 23$ ii. $4g(3.4)$

f. If the square's side length s increases at a constant rate of 3 inches per second, where t represents the number of seconds since the square's side length started increasing, what does $3t$ represent?

g. What does $(3t)^2$ represent in this situation? Expand the expression $(3t)^2$ and explain what this new expression represents and how it compares to the expression $(3t)^2$.

2. The given table represents the retail price (in dollars) of a brand-new Toyota Camry as a function of the number of years, n, after 2000.
   a. What is the input quantity? What is the output quantity?

# of years since 2000, n	Price of car, $f(n)$
0	17,520
2	18,970
3	19,045
5	19,295
6	19,545
7	19,925
10	20,835
13	21,985
20	25,925

b. Evaluate each of the following and describe the meaning in this context.
   i. $f(7)$ ii. $f(20)$

c. Solve for n when $f(n) = 19295$ and explain what your solution represents.

d. Evaluate each expression and describe the meaning in this context.
   i. $f(6) - f(3)$ ii. $f(3) + 1250$ iii. $\frac{f(6)}{f(5)}$

*3. A tortoise and hare are competing in a race around a 400-meter track. The arrogant hare decides to let the tortoise have a 225-meter head start. When the start gun is fired, the hare begins running at a constant speed of 5.5 meters per second and the tortoise begins crawling at constant speed of 2.0 meters per second. Let t represent the number of seconds since the start gun was fired. We are interested in determining who won the race.

   a.   Illustrate the situation with a diagram (mark off the length of the race, note the approximate position of the tortoise from the starting line and the position of the hare, illustrate varying quantities with vectors, etc.).

   b.   Define a function g to represent the tortoise's distance from the finish line in terms of the number of seconds t since the start gun was fired.

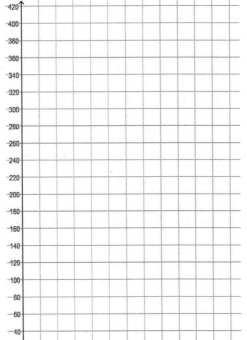

   c.   Define a function h to represent the hare's distance from the finish line in terms of the number of seconds t since the start gun was fired.

   d.   Evaluate $g(40)$ and say what your answer represents.

   e.   Solve $g(t) = 80$ for t and say what the solution represents.

   f.   Solve $h(t) = 0$ and say what the solution represents.

   g.   Construct a graph of g and h on the axes.

   h.   Evaluate $h(20) - g(20)$ and indicate how this quantity is represented on the graph in part (g). What does the quantity represent in the context of this situation?

   i.   Define a new function f where $f(t) = h(t) - g(t)$, and graph this function on the axes in part (g). What does the function f represent? How does the graph of f relate to the graphs of h and g?

j.   What is the slope of the graph of f? What does this value represent in the context of this situation? How does the distance between the tortoise and hare change throughout the race?

k.   Solve $g(t) = h(t)$ for t and say what the solution represents.

l.   Who won the race? Justify your answer using multiple methods.

4.   A hose is used to fill an empty wading pool. The graph shows the volume of the water in the pool (in gallons) as a function of the elapsed time (in minutes) since the pool started filling.
     a.   Define a function g that expresses the volume of water (in gallons) in the pool as a function of t, the elapsed time (in minutes) since the pool started filling.

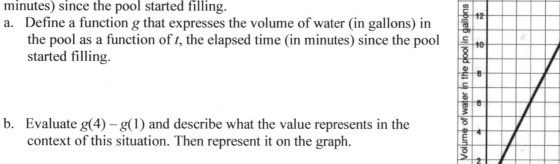

     b.   Evaluate $g(4) - g(1)$ and describe what the value represents in the context of this situation. Then represent it on the graph.

     c.   What could $g(4) + 100$ represent in this context?

*5. As Pat was driving across the flat plains of Kansas with his cruise control on, his gas gauge broke. At the moment the gauge broke, he had 8 gallons of gas in the car's gas tank and his gas mileage was 32 miles per gallon (assume that he maintains this gas mileage by leaving cruise control on). Pat needs to keep track of how much gas is left in his gas tank.
     a.   How many gallons remain in Pat's gas tank after he has driven:
          i. 84 miles since his gas gauge broke?   ii. 150 miles since his gas gauge broke?

          iii. x miles since his gas gauge broke?

     b.   If x represents the number of miles Pat has driven since his gas gauge broke, define a function f to determine the number of gallons left in Pat's gas tank $f(x)$ in terms of x.

c. What is the domain of f?

    What is the range of f?

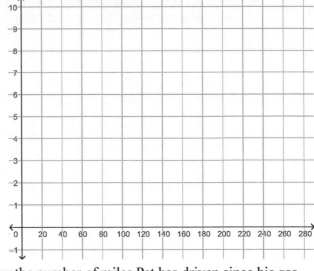

d. What does $f(100)$ represent in the context of this problem?

e. Construct a graph of f on the axes.

f. Explain what the graph of f conveys about how the number of miles Pat has driven since his gas gauge broke, x, and the number of gallons of gas left in Pat's tank, $f(x)$, change together.

g. What is the value of $f(0)$ and what does this value represent in the context of this situation?

h. What is the value of x when $f(x) = 0$ and what does this value represent in the problem context? What point on the graph of f corresponds to where $f(x) = 0$?

i. What are the maximum and minimum values that $f(x)$ can assume in the context of this situation? Explain. (*Note: The maximum value that $f(x)$ can be is also called the maximum value of the function f.*) How do these values compare to f's range?

j. What is the horizontal intercept (zero, root) of f and what does this value represent in the context of this situation?

*1. Running is a popular form of exercise to burn calories and stay healthy. The number of calories burned while running depends on many factors but averages about 100 calories per mile. Suppose Nikki goes for a run, traveling at a constant speed of 720 feet per minute and burning 100 calories per mile she runs.

   a. What quantities are varying (changing) in this situation? What quantities are constant?

   b. Describe how you think each of the following pairs of quantities are changing together.
      i. As the time (in minutes) spent running increases, how does the distance (in feet) Nikki has traveled change?

      ii. As the distance (in feet) Nikki has traveled increases, how does the number of calories she has burned change?

      iii. As the time (in minutes) spent running increases, how does the number of calories Nikki has burned change?

   c. Complete the table of values for this situation. (*Recall there are 5,280 feet in one mile.*)

Time (in minutes) Nikki has been running	Distance (in feet) Nikki has traveled	Distance (in miles) Nikki has traveled
3		
14		
22.2		
t		

   d. How many calories does Nikki burn if she runs 3 miles? 5.7 miles? m miles?

   e. Complete the table of values relating the number of minutes Nikki spends running and the number of calories she burns.

Time (in minutes) Nikki has been running	Number of calories Nikki has burned
4	
11	
19.2	
t	

There are two key takeaways from Exercise #1.

    (i) The expressions we write should be meaningful and related to our understanding of the quantities. For example, note the meaning of each of the following expressions and how new meanings develop as we modify them.

- t represents the number of minutes Nikki has been running
- $720t$ represents the distance Nikki runs (in feet) in t minutes
- $\frac{720t}{5,280}$ represents the distance Nikki runs (in miles) in t minutes
- $100\left(\frac{720t}{5,280}\right)$ represents the number of calories Nikki burns in t minutes

    (ii) Using the results of one calculation or set of calculations as the basis for another calculation is one of the foundations for understanding *function composition* – the process of chaining together multiple function processes.

---

*2. a. Draw a diagram of a square and label the side lengths x (measured in inches). Visualize the square growing and shrinking as the value of x changes and think about how the square's perimeter length (in inches) compares to the side length.

    b. Define a function g that inputs the square's perimeter P and outputs the square's side length x (both measured in inches).

    c. How does the square's side length change as the perimeter changes from 6 inches to 20 inches? Calculate this value **and** represent it using function notation.

    d. Define a function h that inputs the square's side length, x, (in inches) and outputs the square's area, A (in square inches).

    e. Using the functions in parts (b) and (d), determine the area of squares that have each of the following perimeters: 24 inches, 60 inches, and P inches.

Reflecting on Exercise #2 you should notice that we defined two functions and that the output quantity of one function matched the input quantity of the second function. Thus, when given a perimeter, we follow a two-step process.

- **Step 1:** Use *g*, which inputs the perimeter length and outputs the side length (both in inches).
- **Step 2:** Use *h*, which inputs that side length (in inches) and outputs the square's area (in in²).

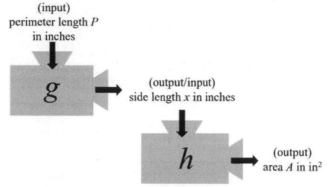

Now, suppose we want to combine these into a single function that combines the two processes together. **That is, we want to define a function *f* that inputs the perimeter length and outputs the area.**

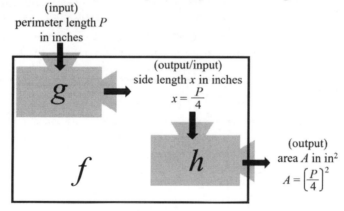

Let's write a formula defining *f*. We'll start with descriptions and move towards a symbolic representation.

$f(\text{perimeter length}) = \text{area}$

$f(\text{perimeter length}) = (\text{side length})^2$ • a square's area is product of its length and width

$f(\text{perimeter length}) = \left(\frac{\text{perimeter length}}{4}\right)^2$ • a square's side length is $\frac{1}{4}$ of its perimeter length

$$\boxed{f(P) = \left(\frac{P}{4}\right)^2}$$

So $f(P) = \left(\frac{P}{4}\right)^2$ with $A = f(P)$. Function *f* is a ***composite function*** – it is formed by uniting the process of two functions to form a single function.

Function Composition

Function composition is the process of chaining together two (or more) function processes by linking the output quantity of one function with the input quantity of another function.

The function created by doing this is called a ***composite function***.

---

3. Using the context in Exercise #2, imagine that the square's perimeter begins at 0 inches and increases at a rate of 3 inches per second.

   a. What is the square's area 8 seconds after the perimeter started increasing? 10 seconds? 40 seconds?

   b. Define a function that inputs the time elapsed (in seconds) since the perimeter started increasing and outputs the square's area (in square inches).

   c. Discuss with a group (or as a class) how the process of using the elapsed time to determine the square's area involves thinking about function composition.

*4. The following graphs show two functions, *f* and *g*. Function *g* takes as its input the high temperature in degrees Fahrenheit and outputs the expected attendance at a neighborhood carnival. Function *f* takes as its input a number of people attending the carnival and outputs the total expected revenue earned by the carnival.

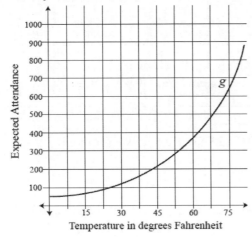
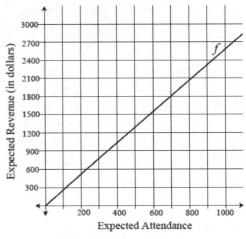

   a. If the forecast predicts clear skies and a high temperature of 45°F, what is the expected revenue from the carnival today? What if the forecast predicts a high temperature of 75°F?

   b. Function *h* inputs the high temperature in degrees Fahrenheit and outputs the expected revenue (in dollars). On the given axes plot at least 5 coordinate points representing input/output pairs for *h*.

In Exercise #4 we used outputs of *g* as inputs to *f* in order to determine the expected revenue given a forecasted high temperature. The notation for representing **expected revenue** given the forecasted high temperature should seem very logical.

expected revenue

$f(\text{expected attendance})$ • *f* inputs expected attendance and outputs expected revenue

$f(g(\text{high temp °F}))$ • $g(\text{high temp °F})$ represents the expected attendance given the high temp °F

$f(g(x))$ • Let *x* represent the high temperature in °F

So $f(g(x))$ represents the expected revenue (in dollars) given *x*, the high temperature in degrees Fahrenheit. Evaluating an expression like $f(g(15))$ is similar to following the order of operations. We begin with the inside function. *We used the graph to estimate each value.*

$f(g(15))$

$f(75)$ • the expected attendance is 75 when the high temperature is 15°F

200   • the expected revenue is $200 when the expected attendance is 75 people

Note that $f(g(x))$ represents the function's outputs, but it is not the function's *name* (just like $f(x)$ is the output of *f*, not the function name). We use the notation $f \circ g$ to name the composite function.

Function Composition Notation

If *f* and *g* are functions, then $f \circ g$ is the name of the composite function formed by chaining together the two processes where *g* is the "inside" function, meaning the process involves:
1.   Inputting a value into *g* and producing an output value.
2.   Using that output value of *g* as an input to *f* to get another output value. This is the output value for the composite function.

If $f \circ g$ is the name of the composite function, then $f(g(x))$ represents the function's output values. *Note that we sometimes condense the notation by just giving this new composite function a name like h. So we could define a function h to be the composite function $f \circ g$ by saying $h(x) = f(g(x))$.*

---

*5. a. In the context from Exercise #4 explain what $f(g(70))$ represents.

b. In the context from Exercise #4 explain why $g(f(70))$ does not represent a real-world quantity.

c. Let's define a new function *k* that is the composition of *f* and *g*, that is, $k(x) = f(g(x))$. Explain what the equation $1800 = k(x)$ represents, then explain how you can find the value of *x* that satisfies the equation.

6. Use the table of values to evaluate or solve each of the following.
    *a. $g(f(-1))$ *b. $f(g(3))$

x	f(x)	g(x)
−2	0	5
−1	3	3
0	4	2
1	−1	1
2	6	−1
3	−2	0

    c. $f(f(3))$ d. $g(g(0))$

    e. If $f(g(x))=3$, then what must be the value of x?

7. Use the graph provided to evaluate or solve each of the following.
    Approximations are acceptable.
    *a. $f(g(2.5))$ *b. $g(f(4))$ c. $g(g(3.5))$

    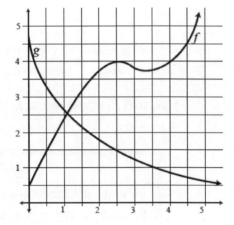

    d. Determine the value(s) of x such that $f(g(x))=2.5$.

8. Use the following functions to answer the questions: $f(x)=\sqrt{x+3}$, $g(x)=2x+9$, $h(x)=\frac{x}{4}$.
    *a. Evaluate $f(g(2))$.

    b. Evaluate $h(f(61))$.

    *c. Function m is defined as $m(x)=g(h(x))$. Write the formula for m.

    d. Function p is defined as $p(x)=g(f(x))$. Write the formula for p.

*1. A pebble is thrown into a lake and the splash creates a circular ripple. The radius length of the circular ripple increases at a constant rate of 7 cm per second. Your goal is to determine the area (in square centimeters) inside the ripple in terms of the number of seconds since the pebble hit the water.

    a. Draw a picture of the situation and label the quantities. Imagine how the quantities are changing together. Discuss in your groups what processes need to be carried out to determine the area inside the ripple when the number of seconds since the pebble hit the water is known.

    b. What quantities are varying (changing) in the situation and how are they changing together?
       i. As the time since the pebble hit the water increases, how does the radius of the ripple change?

       ii. As the radius of the circular ripple increases, how does the area of the ripple change?

       iii. As the time since the pebble hit the water increases, how does the area of the circular ripple change?

    c. Define a function f that determines the radius (in centimeters) of the circular ripple in terms of the number of seconds, t, since the pebble hit the water.

    d. Define a function g to determine the area (in square centimeters) of the circular ripple in terms of the circle's radius length, r (in centimeters).

    e. Describe the process of evaluating $g(f(4.5))$.

    f. In the expression $g(f(4.5))$, describe what quantity and unit of measurement is associated with each of the following.

	Quantity	Unit of measurement
i) 4.5		
ii) $f(4.5)$		
iii) $g(f(4.5))$		

g. Define a function h that determines the area of a circular ripple (measured in square centimeters) as a function of the time t (measured in seconds) since the pebble hit the water.

h. Compute the value of $h(4)$ and the value of $g(f(4))$. What do you observe?

*2. a. If a second pebble hits the water and the radius of the circular ripple increases 3 cm per second (instead of 7 cm per second), write an ***expression*** that represents the varying value of the circle's area in terms of t, the number of seconds since the pebble hit the water.

b. A third pebble hits the water and the radius of the circular ripple increases 5 cm per second. Define a function j that determines the area of the circle in terms of the number of seconds, t, since the pebble hit the water.

*3. a. What is the area of a circle with a circumference that is 15 feet? Describe the multiple step process you used to determine your answer.

b. Define a function h that determines a circle's area in terms of its circumference.

c. Use function h defined in part (b) to determine a circle's area when its circumference is 15 feet. How does your answer compare to your answer in part (a)?

*4. The following graphs show two functions, *f* and *g*. Function *g* takes as its input the radius length of a sphere (in cm) and outputs the sphere's volume (in cm³). The function *f* takes as its input the radius length of a sphere (in cm) and outputs the sphere's surface area (in cm²).

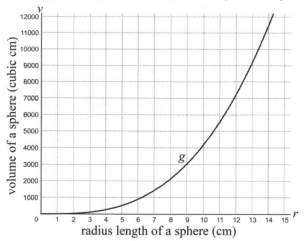

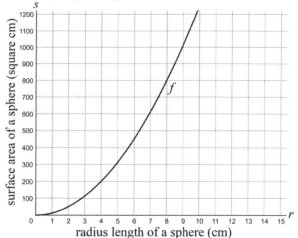

a. Does the expression $f(g(4))$ have a real-world meaning in this context? If so, estimate its value and explain what the value represents. If not, explain your reasoning.

b. Does the expression $g(f(4))$ have a real-world meaning in this context? If so, estimate its value and explain what the value represents. If not, explain your reasoning.

c. Is it possible, using the graphs provided, to determine the approximate volume of a sphere if its surface area is 500 cm²? If so, explain how. If not, explain your reasoning.

d. Is it possible, using the graphs provided, to determine the approximate surface area of a sphere if its volume is 3,000 cm³? If so, explain how. If not, explain your reasoning.

5. A farmer has 250 feet of fencing to create a rectangular pen.
   a. Draw a picture of the situation and label the relevant quantities.

   b. If w represents the pen's width (in feet) and l represents the pen's length (in feet), then the pen's perimeter length is $2w + 2l$ feet, which must always total 250 feet (that is, $2w + 2l = 250$). Solve this equation for w.

   c. If the pen is 85 feet long, what is its width? What is its area? What if the pen is 46.5 feet long?

   d. For any value of w, what must be the corresponding value of l? The corresponding area?

   e. Define a function f that expresses the area of the rectangular pen (in square feet) as a function of w, the width of the rectangular pen (measured in feet).

6. Use the following functions to answer the questions: $f(x) = \sqrt{x+3}$, $g(x) = 2x + 9$, $h(x) = \frac{x}{4}$.

   *a. Evaluate $f(g(2))$.

   b. Evaluate $h(f(61))$.

*7. Let $w = 312q - 100$ and $b = 21q^2$. Define a function h that determines w in terms of b. *Assume that b and q must be positive numbers.*

*1. The formula that determines the perimeter of a square P (in inches) when the square's side length s (in inches) is known, is $P = 4s$.
   a. What thinking is involved in determining the side length, s, when the $P = 20$?

   b. When we solve the equation $20 = 4s$ for s, this is one instance of reversing the process of this formula. Use this same process to determine the side length of a square s (in inches) given the following values of a square's perimeter, P (in inches).
      i. $P = 36$ ii. $P = 22$ iii. $P = 100$

   c. What process (operations) did you use to determine a square's side length when a specific value of a square's perimeter P is known?

   d. Generalize the process you engaged in when answering part (b) by writing a formula that expresses a square's side length s in terms of its perimeter, P.

   d. How is the formula $P = 4s$ and the formula $s = \dfrac{P}{4}$ related? Do they express the same relationship between a square's side length and perimeter? Explain.

We have seen that formulas that produce a unique value of the dependent quantity for every value of the independent quantity are functions and can be represented using function notation. Recall that function notation allows us to reference a specified function relationship by name. We can also discuss a function output (dependent) value such as $f(2)$ without having to determine its value.

If $f(s) = 4s$ then a value of the square's side length (in inches) is input (or put in) to f. Applying the function rule, $4s$, determines a value for the square's perimeter, $f(s)$, in inches.

We can use the image of a "function machine" to evaluate $f(8)$. We input a particular value of 8 in for s, and apply the rule of f (we multiply 4 times 8) to obtain a value for the perimeter, $f(8) = 32$ inches.

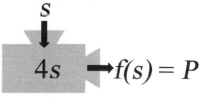

---

*2. Explain how $f(12)$ is processed by f. (*Hint: Use the image above which explains how f processes a value for the input quantity to produce the corresponding value of the output quantity.*)

If $g(P) = \dfrac{P}{4}$, then a value of a square's perimeter (in inches) is input (or put in) to g. Applying the

function rule, $\dfrac{P}{4}$, determines a value of a square's side length, $g(P)$, in inches.

If g is a function that undoes the process of the function f we call the function g the inverse function of f. It is a convention to represent the inverse function of some function f by writing f^{-1}, so we can say that $g = f^{-1}$.

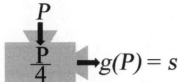

3. a. Evaluate $g(32)$ and use the above image to explain how the value of the input quantity 32 was processed (or operated on) by g.

    b. In groups or as a class, discuss how the function $f(s) = 4s$ and $g(P) = \dfrac{P}{4}$ are related.

    c. What do you notice about how the input and output quantities for f and g are related?

    d. Use function composition to evaluate the following:
       i) $g(f(5))$　　　　　ii) $f(g(25))$　　　　　iii) $g(f(9))$　　　　　iv) $f(g(36))$

    e. T or F: For the functions f and g defined above f undoes the process of g, and g undoes the process of f.

    f. Discuss in your groups (or as a class) how a function f and its inverse function f^{-1} are related in terms of:
       i. the process (or operations) they perform.

       ii. their input quantities (and variables) and output quantities (and variables).

4. When traveling outside the United States it is often useful to be able to determine the temperature in Fahrenheit degrees given the temperature in Celsius degrees. The standard formula for determining the temperature in degrees Fahrenheit F when given the temperature in degrees Celsius C is, $F = \frac{9}{5}C + 32$.

    a. Determine the formula that defines degrees Celsius C in terms of degrees Fahrenheit F. What is the input quantity for this formula? What is the output quantity for this formula?

    b. Define a function g that determines the temperature in degrees Fahrenheit in terms of temperature in degrees Celsius.

    c. Define a function h that determines the temperature in degrees Celsius in terms of the temperature in degrees Fahrenheit.

    d. What does $g(C)$ represent? What does $h(F)$ represent?

    e. Determine the value of:
        i. $g(100)$ ii. $h(212)$ iii. $h\big(g(100)\big)$ iv. $g\big(h(212)\big)$

*5. The function f gives the volume V of a sphere (in cubic inches) as a function of its radius r (in inches). The function formula is $f(r) = \frac{4}{3}\pi r^3$ with $V = f(r)$.

    a. Removing the function notation produces the formula $V = \frac{4}{3}\pi r^3$. Solve this formula for r and explain what this new formula represents.

    b. Let g be the function that inputs a sphere's volume V (in cubic inches) and outputs its radius r (in inches). Use function notation to define a function g that determines a sphere's radius in terms of its volume, V.

    c. Explain what each of the following represents in this context. *You do not need to evaluate them.*
        i. $f(40)$ ii. $g(40)$

        iii. $f(7) \approx 1436.76$ iv. $g(2.14) \approx 0.8$

In Exercise #5 we examined and compared two functions, f and g, that represent the same relationship between two quantities (the radius and volume of a sphere). However, the input and output quantities were switched for the two functions. Thus, representing a volume of 288π (or about 904.8) in^3 for a 6-inch radius is different for each function but the correspondence is the same.

*6. a. Compare and contrast the meaning of $f(6) = 288\pi$ and $g(288\pi) = 6$ given that f and g are defined in Exercise #5.

    b. Determine $g\big(f(6)\big)$ and explain how functions formulas are used to determine this value.

When two functions represent the same relationship between covarying quantities, but their input and output quantities are reversed, we say they are ***inverse functions***. In Exercise #3 the functions f and g are inverses.

The Inverse of a Function

A function and its inverse relation represent the same relationship between two co-varying quantities but with the independent (input) and dependent (output) quantities reversed.

If function g is the inverse of function f, then g undoes the process of f, and f undoes the process of g.

The idea of function inverse shouldn't be entirely new. When you rewrite a two-variable formula to solve for the alternative (or other) variable you are defining the inverse relationship.

7. Solve each of the following formulas for the other variable in the formula.

    *a. $y = 2x - 8$ *b. $w = \dfrac{r + 19}{4}$ c. $n = \dfrac{2x + 6}{x}$

Mathematicians developed a special function name to communicate that two functions are inverses.

Inverse Function Notation

Given a function f, its inverse (if also a function) is named f^{-1}. Since the input and output quantities are reversed in these two functions, we have that, if $y = f(x)$, then $x = f^{-1}(y)$.

The use of "–1" here is not an exponent. It's just part of the function name.

REMINDER: According to the rules of exponents, for any non-zero real number n, $n^{-1} = \frac{1}{n}$. Even though we are using similar notation here, it is ***not true*** that for any function f we have $f^{-1} = \frac{1}{f}$. You have to use context clues to understand the notation and recognize when f^{-1} is referring to the name of a function.

8. Define the inverse function for each of the following functions.

   a. $f(x) = 3x - 7$ with $y = f(x)$

   *b. $j(r) = \frac{2r-5}{6}$ with $w = j(r)$

*9. Use the given table to evaluate the following.

   a. $f(0)$ b. $g^{-1}(3)$ c. $g^{-1}(0)$ d. $f^{-1}(-1)$

x	$f(x)$	$g(x)$
-2	0	5
-1	3	3
0	4	2
1	-1	1
2	6	-1
3	-2	0

Composition of Inverses

Inverse functions represent the same relationship between two quantities but with the input and output quantities reversed. We might say that inverse functions "undo" each other. If $f(a) = b$ then it must be that $f^{-1}(b) = a$ (assuming that f's inverse is a function and assuming that a is an element in f's domain and b an element of its range).

Thus,

$$f^{-1}(f(a)) = f^{-1}(b) \qquad \bullet \text{ since } f(a) = b$$
$$= a \qquad \bullet \text{ since } f^{-1}(b) = a$$

$$f(f^{-1}(b)) = f(a) \qquad \bullet \text{ since } f^{-1}(b) = a$$
$$= b \qquad \bullet \text{ since } f(a) = b$$

The composition of two inverses will always produce the original input value for the "inside" function (in either order). In fact, this is one way to test if two functions are inverses.

10. Your classmate believes that each of the following pairs of functions are inverses (and thus used inverse function notation to define them). Is your classmate correct? Verify by creating the formula for $f^{-1}(f(x))$ and simplifying.

   *a. $f(x) = 2(x+1) - 8$ and $f^{-1}(x) = \frac{x+8}{2} - 1$

   b. $f(x) = \frac{3x+7}{4}$ and $f^{-1}(x) = \frac{4}{3}x - 7$

11. Sarah and Maria are racing. Sarah gives Maria a 30-yard head start and then she starts running. Once Sarah starts running she travels at a constant rate of 4.5 yards per second while Maria travels at a constant rate of 3 yards per second.
    a. Draw a diagram of this situation.

    b. Write a formula for function f that models Maria's lead d (in yards) in terms of t, the number of seconds since Sarah began running.

    c. Write a formula for function f^{-1}.

    d. Create the composite function $f\left(f^{-1}(d)\right)$ to demonstrate that f and f^{-1} are inverses.

    e. How long does it take Sarah to catch up to Maria? Determine the value and then demonstrate how to represent this value using function notation in two different ways.

12. Use the given table to evaluate the following.
    a. $g^{-1}(3)$ b. $g^{-1}(0)$

    c. $f^{-1}(g(2))$ d. $g^{-1}(f^{-1}(6))$

x	$f(x)$	$g(x)$
−2	0	5
−1	3	3
0	4	2
1	−1	1
2	6	−1
3	−2	0

*13. Note that the inverse relation of a function **might not** be a function itself. Given the function f defined by $f(x) = x^2$ with $y = f(x)$, complete the following.

a. Evaluate $f(2)$ and $f(-2)$.

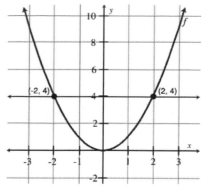

b. Explain why f^{-1}, that is f's inverse function, is not a function. (Recall the definition of a function.)

c. Consider $g(x) = x^2$ where the domain of g is restricted so that $x \geq 0$.

i. Explain why g's inverse is a function.

ii. Write the formula for g^{-1} and give its domain and range.

14. For each of the following functions f determine if its inverse will also be a function (based on the information you know). If not, then choose a restricted domain for f so that its inverse would be a function.

a.

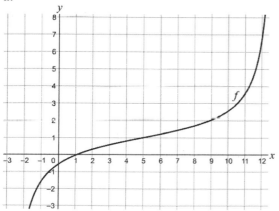

b.

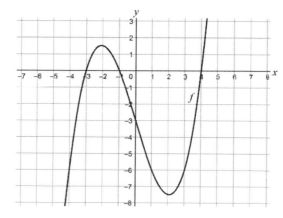

*1. A car is driving away from a crosswalk. The distance d (in feet) of the car from the crosswalk t seconds since the car started moving is given by the function $f(t) = \frac{1}{2}t^2 + 3$ with $d = f(t)$.

   a. As the number of seconds since the car started moving increases from 4 seconds to 9 seconds, what is the change in the car's distance from the crosswalk?

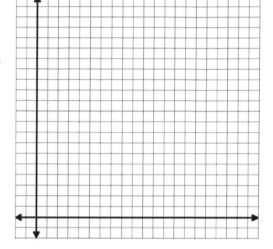

   b. Graph f on the given axes, then illustrate how to represent each of the following.
      i. The increase in t from 4 to 9 seconds.
      ii. The corresponding change in the car's distance from the crosswalk as the value of t increased from 4 to 9 seconds.

   c. True or False: The car travels at a constant speed as the value of t increases from 4 to 9 seconds. Discuss and explain. (*Hint: It may help to think about how far the car travels in the 1ˢᵗ second compared to how far it travels in the 2ⁿᵈ second*).

   d. Since the car is not traveling at a constant speed it can be challenging to estimate its speed during the interval. It is a common practice to <u>estimate</u> the speed of the car during the interval by pretending its speed was constant and asking, "What constant speed would have resulted in the car traveling the same distance (as the distance that the car actually traveled) over this time interval from $t = 4$ to $t = 9$ seconds?"
      i. Plot the points $\left(4,\, f(4)\right)$ and $\left(9,\, f(9)\right)$ on the graph and draw the line passing through them.
      ii. Determine the constant rate of change for the relationship you just drew.

      iii. <u>If</u> a car was traveling at the speed you found in part (ii), how far would it travel as t changes from $t = 4$ to $t = 9$?

   e. True or False: The line represents the graph of the actual distance (in feet) of the car from the crosswalk as t increases from 4 to 9 seconds. Explain.

As we mentioned, it's common to substitute the behavior of a linear function over an interval as a way to approximate the behavior of other functions that don't have a constant rate of change. The constant rate of change we get by doing this is called the ***average rate of change*** of the function over that interval.

<u>Average Rate of Change</u>

The average rate of change of a function on an interval from $x = a$ to $x = b$ is the constant rate of change that would have the same net change in output values over that interval.

---

*2. Given $f(x) = \sqrt{x+3}$ and $g(x) = \frac{10}{x}$, do the following.

   a. Find the average rate of change of f over the interval from $x = 1$ to $x = 6$ and then again over the interval from $x = 3$ to $x = 10$. Explain what your solutions represent.

   b. Find the average rate of change of g over the interval from $x = 1$ to $x = 4$ and then again over the interval from $x = 5$ to $x = 10$. Explain what your solutions represent.

*3. Use function notation to represent the average rate of change of a function f over the interval from $x = 4$ to $x = 7$.

*4. Use function notation to represent the average rate of change of a function g over the interval from $x = -3$ to $x = 0$.

*5. Use function notation to represent the average rate of change of a function h over the interval from $x = n$ to $x = n + 2$.

The expression that represents or calculates the average rate of change is called **the difference quotient**.

The Difference Quotient

The average rate of change of a function on any interval h units long is represented by **the difference quotient**:

$$\frac{f(x+h)-f(x)}{(x+h)-x} \quad \text{or} \quad \frac{f(x+h)-f(x)}{h}$$

---

*6. Use the difference quotient to represent the average rate of change of the car's distance from the cross walk in terms of the time elapsed from Exercise #1.

I. THE BOX PROBLEM AND MODELING RELATIONSHIPS (TEXT: S1, S2)

1. A box designer has been charged with the task of determining the volume of various boxes that can be constructed by cutting four equal-sized square corners from a 14-inch by 17-inch sheet of cardboard and folding up the sides.
   a. What quantities vary in this situation? What quantities remain constant?
   b. Create an illustration to represent the situation and label the relevant quantities.
   c. What are the dimensions of the box (length, width, and height) if the length of the side of the square cutout is 0.5 inches? 1 inch? 2 inches?
   d. Define a formula to relate the height of the box and the length of the side of the square cutout. Be sure to define your variables.
   e. Let x represent the length of the side of the square cutout; let w represent the width of the base of the box; let l represent the length of the base of the box; and let V represent the volume of the box. Complete the given table assuming each quantity is measured in a number of inches (volume measured in cubic inches).

x	w	l	V
0	14	17	0
4			
9			
		2	
	0		

   f. Define a formula to relate the length of the base of the box and the length of the side of the square cutout. Be sure to define your variables.
   g. Define a formula to relate the volume of the box to the length of the side of the square cutout.
   h. Use the formula created in part (h) to determine the change in the volume of the box as the length of the side of the square cutout increases from 0.5 inches to 1 inch and from 1 inch to 1.5 inches. Why are these values not the same?
   i. Use your graphing calculator to approximate the maximum volume of the box rounded to the nearest tenth of a cubic inch. Explain how you know that the value you obtained is the maximum value of the volume of the box rounded to the nearest tenth.

2. A box designer has been charged with the task of determining the surface area of various open boxes (no lids) that can be constructed by cutting four equal-sized surface corners from an 8-inch by 11.5-inch sheet of cardboard and folding up the sides. (*Note: The surface area is the total area of the box's sides and bottom.*)
   a. What quantities vary in this situation? What quantities remain constant?
   b. Create an illustration to represent the situation and label the relevant quantities.
   c. Complete the given table.

Cutout length (in)	Height of box (in)	Length of box (in)	Width of box (in)
0.5			
1			
2			

   d. Define a formula for the following relationships:
      i. the length of the base of the box in terms of the length of the side of the square cutout
      ii. the width of the box in terms of the length of the side of the square cutout
   e. Define a formula that relates the total surface area, s, (measured in square inches) of the open box to the size of the square cutout x (measured in inches).
   f. There are two ways to think of determining a box's surface area: by (1) adding up the area of each piece of the box; and by (2) subtracting the area of the four cutouts from the area of the initial sheet of cardboard.
      i. Which method did you use in defining your formula in part (e)?
      ii. Define a formula to compute the box's surface area by the other method. Compare the two formulas. Are they equivalent (that is, do they always produce the same result)? Show this algebraically.

3. An open box is constructed by cutting four equal-sized square corners from a 10-inch by 13-inch sheet of cardboard and folding up the sides.
    a. Define a formula to relate the volume of the box to the length of the side of the square cutout. Be sure to define your variables.
    b. Determine the volume of the box when the length of the side of the square cutout is:
        i. 1 inch   ii. 2.5 inches   iii. 5 inches
    c. Use the graph to approximate the value(s) of the length of the side of the square cutout when the volume is:
        i. 0 in³   ii. 100 in³   iii. 108.4 in³
    d. As the length of the side of the square cutout increases from 0 to 1 inch, by how much does the volume change? Illustrate this change on the graph.
    e. For what interval(s) of the cutout length is the box's volume increasing? Decreasing?

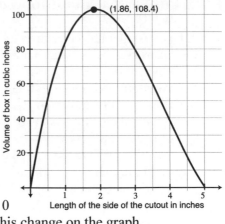

For Exercises #4-5 use the following context. *Windows, Inc. manufactures specialty windows. One of their styles is in the shape of a semicircle as shown.*

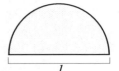

4. When a customer orders this window, they specify the length of the base of the window.
    a. Determine the total length of frame needed for a window with a base of 4 feet.
    b. Determine a formula that relates the total length of the frame to the length of the base of the window. Be sure to define your variables.
    c. If the cost of the framing material is $12 per linear foot, what is the cost of a window frame with a base-length of 4 feet?
    d. Suppose that your budget limits you to spending $500 on your window frame. What is the longest base length that you can afford?

5. Suppose that the glass pane costs $23 per square feet.
    a. What is the area of a glass pane when the base of the window is 6 feet?
    b. Write a formula that relates the area of the glass pane to the length of the base of the window. Be sure to define your variables.
    c. What is the cost of the glass pane when the base of the window is 6 feet?
    d. Suppose that your budget limits you to spending $850 on your glass pane. What is the longest base length of the window that you can afford?
    e. The total cost of the window includes both the cost of the glass pane at $23 per square foot and the cost of the window frame at $12 per linear foot. What is the total cost of a window with a base of 3 feet? (*Hint: Consider the formulas you defined in Exercises 4 and 5.*)

6. Find the value of $\dfrac{(9+x)-x^2}{2x+7}$ when $x = 3$.

7. Find the value of $\dfrac{3x^2 + x - 2(3x+5)}{2x+9}$ when $x = 8$.

8. Find the value of $\dfrac{\frac{30}{y} + \frac{49}{y+2} + \frac{100}{y^2}}{2y}$ when $y = 5$.

9. Find the value of $\dfrac{(x+4)^3 - 4y + 6xy}{-5x+6y}$ when $x = 2$ and $y = 0.5$.

10. Find the value of $y\sqrt{x+4} - 3x - 7y + \frac{4x}{8}$ when $x = 12$ and $y = -4$.

11. Find the value of $\left(2(x-7)\right)^2 - \frac{y}{6}$ when $x = -2$ and $y = 54$.

II. FUNCTION RELATIONS, DOMAIN AND RANGE, AND NOTATION (TEXT: S1)

For Exercises #12-16, do the following.
   a. Determine the input quantity,
   b. Determine the output quantity,
   c. Determine if the relationship is a function. If it is, describe a possible domain and range.

12. A student's ID number with respect to the year of a student's birth.

13. The year of a student's birth in terms of the student's ID number

14. In a given apartment complex, the apartment number in terms of the number of people living in that apartment.

15. In a given apartment complex, the number of people living in that apartment in terms of the apartment number.

16. According to the information in the table, is the height of a person in terms of his/her name a function relationship? Explain your reasoning.

17. According to the information in the table, is the name of the person in terms of his/her height a function relationship? Explain your reasoning.

Person's Name	John	Mary	Sally	Michael	Dawn
Person's Height (inches)	71	63	64.5	71	68

18. For each of the given relations, determine whether y is a function of x based on the given information. Use the definition of a function to justify your answer.

   a.

x	y
1	3
2	2
1.7	4.5
2.1	9
2	1.1
5	7

   b.

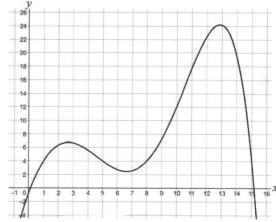

   c. $y = x(8.5 - 2x)(11 - 2x)$

   d. $y = 2^x$

   e. A manufacturing plant produces bags of plaster weighing between 0 and 90 pounds. Bags weighing up to 50 pounds are marked with a "1" (indicating that the bags can safely be carried by one person), and bags weighing more than 50 pounds are marked with a "2" (indicating that the bags are heavy and should be carried by two people). Is the number marked on the bag y a function of the weight of the bag x?

19. For each of the relations in Exercise #18, determine whether x is a function of y. Justify your answers.

20. Create two tables that represent relations between two variables that are functions. Explain why one variable is a function of the other variable.

21. Create two tables that represent relations between two variables that are NOT functions. Explain why the relationships are not functions.

Use the following functions for Exercises #22-26.

$$f(x) = 3x^2 + 5x - 7 \qquad h(x) = \frac{x}{5x - 10} \qquad k(x) = \sqrt{x + 2} \qquad n(x) = \frac{\sqrt{x - 7}}{x^2 - 4}$$

22. Evaluate each of the following expressions.
    a. $f(6)$ b. $g(-2)$ c. $k(9)$ d. $n(11)$

23. Explain the meaning of each of the following statements.
    a. $f(-3) = 5$ b. $k(17) \approx 4.36$

24. a. Explain what it means to find x such that $k(x) = 5$. (*Don't determine its value yet.*)
    b. Find the value(s) of x such that $k(x) = 5$.

25. Determine the domain of functions f and h.   26. Determine the domain of functions k and n.

Use the graphs of f and g for Exercises #27-31.

27. Evaluate each of the following expressions. *Approximations are okay.*
    a. $f(2)$ b. $f(0)$ c. $g(0)$ d. $g(4.5)$

28. Explain the meaning of each of the following statements.
    a. $f(1) \approx 0.24$ b. $g(2.5) \approx -3$

29. a. Explain what it means to find x such that $g(x) = 4$. (Don't determine its value yet.)
    b. Find the approximate value(s) of x such that $g(x) = 4$.

30. Determine the range of f (based on the information you know from the graph).

31. Determine the range of g (based on the information you know from the graph).

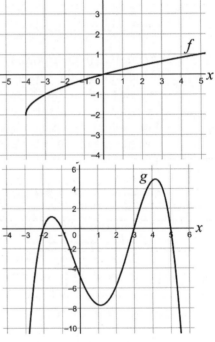

32. The expression $224 - 4x^2$ calculates the surface area of a box (measured in square inches) when the box is made with a square cutout of side length x (measured in inches) from a 14" by 16" sheet of paper.
    a. Define a function g that has the length of the square cutout x as input (measured in inches) and the surface area A of the corresponding box as output (measured in square inches).
    b. Using function notation, represent the surface area of the box when the length of the side of the square cutout is 0.2 inches, 2.7 inches, and 4.1 inches. (That is, *do not* calculate the surface area. Instead *represent* the surface area determined by each length of the side of the square cutout.)
    c. What does the expression $g(1.3)$ represent?

33. The expression $\frac{4}{3}\pi r^3$ calculates the volume of a sphere (measured in cubic inches) when the radius of the sphere is r inches.
    a. Define a function h that has the radius of the sphere r as input (measured in inches) and the volume of the sphere V as output (measured in cubic inches).   *Exercise continues on the next page.*

b. Using function notation represent the volume of the sphere when the radius of the sphere is 2.4 inches, 3.1 inches, and 5.2 inches. (That is, *do not* calculate the volume. Instead *represent* the volume determined by each radius length.)

c. What does the expression $h(19.6)$ represent?

III. Using and Interpreting Function Notation (Text: S3, S4)

34. The expression $224 - 4x^2$ calculates the surface area of a box (measured in square inches) when the box is made with a square cutout of side length x (measured in inches) from a 14" by 16" sheet of paper. Let g be a function that takes as its input the length of the side of the square cutout (measured in inches) and outputs the surface area of the corresponding box (measured in square inches).
    a. What does the expression $g(1.5) - g(0.5)$ <u>represent</u> in this context? What is its value?
    b. What does the expression $g(6.25) - g(2.1)$ <u>represent</u> in this context? What is its value?

35. Functions are commonly used by computer programmers to access databases. All databases that involve time have "reference times" (times they consider to be "0"). A certain database uses a function named g to access the population of Loveland, CO and uses January 1, 1990 as its reference time. So, $P = g(u)$ represents Loveland's population u years from January 1, 1990.
    a. Represent Loveland's population on 1/1/1982.
    b. Represent the change in Loveland's population from 1/1/1994 to 1/1/1997.
    c. Represent a population value 6 times as large as Loveland's population on 1/1/1972.
    d. What does $g(8) - g(-13)$ represent?
    e. What does $g(h + 3)$ represent, where h is some number of years?
    f. What does $g(h + 3) - g(3)$ represent, where h is some number of years?

36. Function h determines a dairy farmer's cost (in dollars) to produce b gallons of milk. Explain what each of the following represents in this context.
    a. $h(18)$ b. $h(b) = 25$ c. $h(15) - h(9)$ d. $h(4.3) + h(6.4)$

37. Use function notation to represent each of the following.
    a. When the input to the function q is 10, the output is 12.
    b. The output of a function h is c when the input value is 34.
    c. Function f expresses the area of a square as a function of its side length s.
        ii.   The area of a square is 25m^2 when the side length is 5m.
        iii. Write an expression using the function f that represents how much larger the area of a square with side length of 6 inches is than the area of a square with side length of 2 inches.
    d. The circumference of a circle is π times as large as the circle's diameter d.
        i.   Define a function g to express the circumference of the circle in terms of its diameter.
        ii. Represent the circumference of a circle with a diameter length 3 times as large as a inches.

38. The given table provides values of the circumference length C of a circle (in cm) as a function of its radius length r (in cm).
    a. What does the expression $f(1.5)$ represent? What is the numerical value of $f(1.5)$?
    b. What does the expression $f(2) - f(1.5)$ represent? What is the numerical value of $f(2) - f(1.5)$?
    c. When the radius of a circle changes from 1.5 to 2.5 cm, by how much does the circle's circumference change?
    d. Express the solution to part (c) using function notation.

radius length (cm) r	circumference (cm) $C = f(r)$
1	6.283
1.5	9.425
2	12.566
2.5	15.708
3	18.850

39. Given the existence of some function *b* and *k*:
    a. Use function notation to represent the change in the output value of function *k* when the input value changes from 6 to 9.
    b. Use function notation to represent the sum of the output of the function *k* at an input of –9 and the output of the function *b* at an input of 6.12

40. Use the following information to answer the questions below.
    - *C(t)* represents the number of cats owned by people living in the U.S. *t* years since 2000.
    - *D(t)* represents the number of dogs owned by people living in the U.S. *t* years since 2000.
    a. Represent the total number of cats and dogs *P(t)* owned by people in the U.S. *t* years since 2000.
    b. Represent, using function notation, how many _more_ cats than dogs were owned by people living in the U.S. *t* years since 2000.

41. Evaluate each of the following:

    a. $f(13)$ when $f(x) = 4.5 - 6x$

    b. $f(5)$ when $f(x) = \dfrac{2x^2 + (6 - 4x)}{9 - 3x}$

    c. $g(2.6)$ when $g(x) = \dfrac{x^3 - 2.7x - 5(-2x + 4)}{-5.9 + 3.2x}$

    d. $m(-6.1)$ when $m(x) = \dfrac{(x + 4)^2 - (4.1x)^2 + 3x}{1.1 + 7x}$

42. Evaluate each of the following:

    a. $f(x + 2)$ when $f(x) = 4x^2 - 2x + 10$

    b. $h(2x)$ when $h(y) = \dfrac{y^3 - 2y^2 + 4}{2y}$

IV. FUNCTION COMPOSITION: CHAINING TOGETHER TWO FUNCTION PROCESSES (TEXT: S5)

43. Alejandra bought a new house recently and is planning to install landscaping next week. In her initial budget, she set aside $200 to purchase ¾-inch gravel to cover part of her yard. This size gravel covers approximately 130 square feet per ton. She has four different options for this size gravel depending on the quality of the gravel. In addition to the gravel cost the company charges $59 for delivery.

¾-inch gravel type	Cost per ton (in dollars)
Grade A	$31.50
Grade B	$26.50
Grade C	$24.50
Grade D	$22.50

    a. i. How many tons of grade B gravel can Alejandra afford if she must pay the $59 delivery charge?
       ii. How many tons of grade D gravel can Alejandra afford if she must pay the $59 delivery charge?
    b. How many square feet of her yard can she cover with grade C gravel? Explain your thinking.
    c. How many square feet of her yard can she cover with grade A gravel? Grades B, C, or D?

44. Reggie, a college student, works part-time during the school year. He has to budget his expenditures so that he has enough money at the end of the month to cover his rent, car payment, and insurance. Currently, Reggie can only afford $15 per week for gas.

    a. As the price of gas fluctuates, the amount of gas Reggie can purchase each week varies. Complete the table of values showing the number of gallons of gas Reggie can purchase with $15 at the given fuel prices.

Price of fuel (in dollars per gallon)	Number of gallons of fuel Reggie can purchase for $15
3.199	
3.499	
3.599	
3.799	
p	

    b. Reggie's car gets an average of 28 miles per gallon. How many miles can he drive in a week if he purchases 3 gallons of fuel? 4.18 gallons? *g* gallons?

Exercise continues on the next page.

c. Explain how you can determine the number of miles Reggie can drive in a week if gas costs $3.899 per gallon and he has $15 to spend on gas.

d. Complete the table of values relating the price of fuel (in dollars per gallon) and the number of miles Reggie can drive on $15 worth of gas.

e. How many miles can Reggie drive on $15 worth of gas if gas costs p dollars per gallon?

Price of fuel (in dollars per gallon)	Number of miles Reggie can drive on $15 worth of gas
3.299	
3.449	
3.579	
3.839	

45. Jessie does a lot of traveling for business. When he travels he likes to go for a run in the morning to keep fit. Jessie has noticed that the elevation of the city he is visiting impacts how long he is able to spend running since there is less oxygen available at higher elevations. The graphs below provide information about Jessie's exercise routines recorded over many business trips.

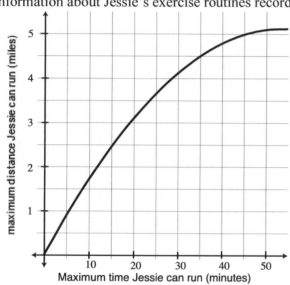 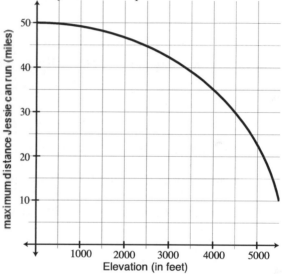

a. Describe how to determine the maximum distance Jessie can run if he is visiting a city with an elevation of 4000 feet.

b. Explain how you can determine the elevation of the city Jessie is visiting if he expects his maximum running distance to be 5 miles.

46. A ball is thrown into a lake, creating a circular ripple with a radius length that increases 7 cm per second. We want to express the circle's area in terms of the time elapsed since the ball hit the lake.

a. Draw a diagram of the situation.

b. Identify the quantities in the situation whose values vary and state what units you'll use to measure each of these quantities. Repeat for quantities whose values are fixed.

c. As the amount of time t in seconds since the ball hit the lake increases over each of the given time periods, how does the radius r of the ripple (in centimeters) change?
   i. from $t = 0$ to 3 seconds   ii. from $t = 4$ to 6 seconds   iii. from $t = 6$ to 6.5 seconds

d. Define a function g that defines the radius r of the ripple in terms of the time t in seconds since the ball hit the water.

e. Define a functions f that determines the area of the ripple A in terms of the time t in seconds since the ball hit the water.

f. Define a composite function h that expresses the area of the circle as a function of the time elapsed since the ball struck the water.

g. Describe the meaning of $h(2.3)$ without performing any calculations. Then calculate and interpret the meaning of the value of this expression.

47. When hiring a contractor to add insulation to your attic, he provides you with information that you convert into tables specifying how input and output values are related for two functions, f and g.

Function f		Function g	
Number of bags of insulation applied, n	Depth of insulation (inches), $f(n)$	Depth of insulation (inches), d	Estimated annual heating/ cooling costs (dollars), $g(d)$
11	2	15	$940
20	3.6	13	$1,000
28	5	11	$1,100
40	7.2	9	$1,250
50	9	7	$1,500
61	11	5	$1,900
70	12.6	4	$2,750
78	14	3	$3,600

   a. Describe the meaning of $g(11)$.

   b. Does the expression $f(g(11))$ have a real-world meaning in this context? If so, find the value and explain its meaning. If not, explain your reasoning.

   c. Does the expression $g(f(11))$ have a real world-meaning in this context? If so, find the value and explain its meaning. If not, explain your reasoning.

   d. Solve the equation $g(f(n)) = 1900$ for n and explain the meaning of your solution.

   e. Function T is defined as follows: $T(n) = g(f(n))$. Use the diagram to the right to describe what each of the parts (A, B, C, D, and E) of the definition represents.

$$T(n) = g(f(n))$$

48. Suppose you have a coupon for $10 off a purchase of $100 or more at Better Buys.

   a. Define a function C that determines the final cost of a purchase after applying the coupon if x is the original price (in dollars). Explain what the expression $C(215.83)$ represents.

   b. What are the domain and range of the function C?

   c. This weekend only Better Buys has a customer appreciation sale that offers 5% off all purchases. Define a function T that determines the final cost of a customer's purchase after the sale if the original purchase price is p dollars. (Do not include the coupon in this function rule.)

   d. Suppose Better Buys will allow you to use both the sale and the coupon together. If you buy a television set that is priced at $1979.99, describe what the expression $C(T(1979.99))$ represents and determine its value.

49. An oil tanker crashed into a reef off of the coast of Alaska, grounding the tanker and punching a hole in its hull. The tanker's oil radiated out from the tanker in a circular pattern. You are an engineer for the oil company and your task is to monitor the spill.

   a. Draw a diagram of the situation.

   b. What are the varying quantities in this situation? Define variables to represent the values of the varying quantities. Be sure to include the units of measure when defining the variables.

   c. If the oil tanker was 105 feet from the spill's outer edge exactly 24 hours after the spill, how much area (in square miles) did the spill cover at that time.

   d. News reporters want to know how much oil has spilled at any given time. The only information available to you is satellite photos that include the spill's radius (in feet). You also know that 7.5 gallons of oil creates 4 square feet of spill.

      i. Define a function g that expresses the number of gallons of oil spilled, x, in terms of the area of the oil spill, A, (in square feet).

      ii. Define a function, f, that expresses the number of gallons of oil spilled in terms of the spill's radius (in miles).

   e. Use your function from part (c(ii)) to represent the increase in the number of gallons spilled as the radius of the oil spill increases from 121 feet to 152 feet.

50. Given $f(x) = 2x + 1$ and $g(x) = x^2 + 2x + 1$, complete the following.

    a. Fill in the first two tables and use these two tables to complete the third table.

x	$f(x)$
0	
1	
1.5	
2	
5	

x	$g(x)$

x	$g(f(x))$
0	
1	
1.5	
2	
5	

    b. Describe how you used the first two tables to determine the values of the third table.

51. Use the given tables to answer the following questions

    a. Evaluate the following expressions:

        i. $g(f(-2))$ ii. $f(g(0))$

        iii. $g(g(4))$ iv. $f(g(2))$

        v. $g(f(-1))$ vi. $f(f(0))$

    b. Solve the equation $g(f(x)) = -10$ for x.

    c. Solve the equation $f(g(x)) = 16$ for x.

x	$f(x)$
-4	16
-3	13
-2	6
-1	0
0	-4
1	-7
2	-9
3	-10

x	$g(x)$
-8	-20
-6	-17
-4	-10
-2	-4
0	-1
2	2
4	6
6	11

52. Use the words *input* and *output* to express the meaning of the following expressions. Assume that g and h are functions.

    a. $h(g(2))$ b. $g(g(3))$ c. $g(h(x+1))$

53. The functions f, g, and h (given below) are used to define the functions s, r, and v. Re-write the definitions of s, r, and v so that their definitions do not involve function composition.

$$f(x) = \frac{2x+1}{x-3} \qquad g(x) = x^2 - 7 \qquad h(x) = -3x + 2$$

    a. $s(x) = f(g(x))$ b. $r(x) = h(h(x))$ c. $v(x) = g(h(x))$

54. The functions g, h, and k are defined below. Use these functions to answer the questions that follow.

$$g(x) = x + 5$$

x	$h(x)$
-1	-13.4
0	-7.3
2	-4.4
4	3
5	6.8
6	15
8	22

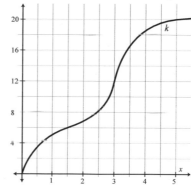

    a. $k(g(0))$ b. $h(k(1.5))$ c. $g(h(2))$ d. $k(h(4))$

    e. Solve the equation $h(g(x)) = 22$ for x.

    f. Solve the equation $g(k(x)) = 17$ for x.

55. Use the graphs below to answer the questions that follow

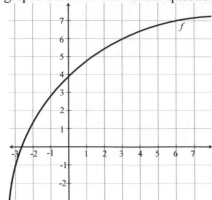 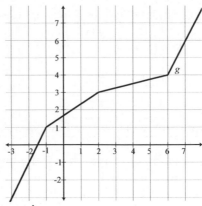

a. Approximate the value of each of the following expressions.

i. $f(f(3))$ ii. $g(f(6))$ iii. $g(g(-2))$ iv. $g(f(-3))$

b. Find the value of x that satisfies each of the following equations.

i. $f(g(x)) = 6$ ii. $g(f(x)) = 4$

56. Functions g and r are defined by their graphs to the right.

a. Determine the values of each of the following expressions:

i. $r(g(2))$ ii. $g(r(1))$ iii. $r(r(6))$ iv. $g(r(-2))$

b. How does the output $g(r(x))$ vary as x varies from 1 to 3?

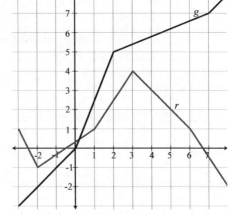

57. For each of the functions defined below, redefine that function in terms of two new functions, f and g, using function composition and function arithmetic. *For example, $f(x) = (2x)^3$ can be defined as $f(x) = h(g(x))$ if $g(x) = 2x$ and $h(x) = x^3$.*

a. $h(x) = 3(x-1)+5$ b. $m(x) = (x+4)^2$

c. $k(x) = (x+2)^2 + 3(x+2) + 1$ d. $j(x) = \sqrt{x-1}$

e. $p(x) = \frac{500}{100-x^2}$

V: Extra Practice with Function Composition (TEXT: S5)

58. The standard formula for determining temperature in degrees Fahrenheit when given the temperature in degrees Celsius is $F = \frac{9}{5}C + 32$. We can write this formula using function notation by letting

$F = g(C)$ and writing $g(C) = \frac{9}{5}C + 32$. The function g defines a process for converting a temperature measure in degrees Celsius to degrees Fahrenheit.

a. State the meaning of $g(100)$ and evaluate $g(100)$.

b. Solve the equation $g(C) = 212$ and explain how you arrived at your answer.

c. Define a function h that converts temperature measures in degrees Fahrenheit to the corresponding measure in degrees Celsius. (*Hint: Generalize the steps you described in part (b) that reversed the process of g. You can also solve $F = \frac{9}{5}C + 32$ for C.*)

d. State the meaning of $h(212)$ and evaluate $h(212)$.

e. Determine the values of $g(h(212))$ and $h(g(100))$ <u>without performing any calculations</u>. What do you notice about the relationship between g and h?   *Exercise continues on the next page.*

f. Represent the value of $g(h(n))$. Represent the value of $h(g(k))$. What do you notice about the relationship between g and h?

g. Use function notation to represent the relationship between functions g and h based on your answers to (e) and (f).

59. A spherical bubble inflates so that its volume increases by a constant rate of 120 cm^3 per second.
   a. Define a function f that expresses the radius of the bubble, r, (in centimeters) as a function of the bubble's volume, V, (in cubic centimeters). (*Note: The volume of a sphere is* $V = \frac{4}{3}\pi r^3$).
   b. Define a function g that expresses the bubble's volume, V, (in cubic centimeters) as a function of the number of seconds, t, since the bubble began to inflate.
   c. Use your functions from (a) and (b) to define a function h that expresses the bubble's radius, r (in centimeters) as a function of the number of seconds, t, since the bubble began to inflate.
   d. Use function notation to represent the change in the bubble's radius as the number of seconds since the ball began to inflate increases from 5 to 5.3 seconds.

60. A farmer decides to build a fence to enclose a rectangular field in which he will plant a crop. He has 1000 feet of fence to use and his goal is to maximize the area of his field.
   a. Draw a diagram that shows the quantities in this situation. Label the changing quantities with variables, and then explain what each variable represents in terms of the situation. (e.g., define each quantity and its unit of measure).
   b. What are the dimensions of the enclosed field if one side must be 200 feet? What are the dimensions of the enclosed field if one side must be 352.41 feet?
   c. Determine a formula that defines the length of the field l in terms of the field's width w given that the total amount of fence is 1000 feet.
   d. Describe how the length of the field changes as the width of the field increases.
   e. Determine a formula that relates the length and width of the field to the total area of the enclosed field.
   f. Using parts (c) and (e), define a function f that expresses the area of the field (measured in square feet) as a function of the length of the side of the field l (measured in feet).
   g. Create a graph that represents the area of the enclosed field in terms of the length of the side of the field. Describe how the area of the enclosed field changes as the length of the side of the field increases.
   h. Based on your graph, what is the maximum area of the enclosed field? What are the dimensions of the enclosed field that create this maximum area?

61. Janet works 36 hours a week between two part-time jobs during the semester: residence hall front desk assistant ($7.50 per hour) and supermarket cashier ($9.25 per hour).
   a. Let r represent the number of hours Janet works in a given week as a residence hall front desk assistant. Write a formula to calculate the value of s, the number of hours Janet works as a supermarket cashier the same week.
   b. Define a function f that calculates the amount of money Janet makes for working r hours as a resident hall front desk assistant and then define a function g that calculates the amount of money Janet makes as a supermarket cashier working r *hours as a residence hall front desk assistant.*
   c. Define a function T that calculates the total amount of money Janet makes in a given week if she is working both jobs and works r hours at the residence hall front desk.
   d. The supermarket requires Janet to work between 15 and 30 hours each week. What is the domain of function T? What is the range of function T?

62. You are planning a trip to Japan where the currency is the yen. On your way to Japan you will stop in Italy where the currency is the euro. You know that you will need to convert your US dollars to euros before your trip and know that the number of euros is 0.78 times as large the number of dollars. You also know the number of yen is 103 times as large as number of dollars.
    a. Write a formula that expresses the number of yen, y, you will have if you begin with d dollars.
    b. Write a formula that expresses the number of euro, e, you will have if you begin with d dollars
    c. You know that you will not be converting from dollars to yen because you will first stop in Italy. So it would be more helpful to know the conversion rate between euros and yen. Write a function g that expresses the number of yen you will have, y, in terms of the number of euros you convert.

63. The length of a steel bar changes as the temperature changes. Consider a 10-meter steel bar that is placed outside in the morning when the temperature is 20 degrees Celsius. Let l represent the length of a steel bar in meters. Let t be the temperature of the steel bar in degrees Celsius. And let n be the number of minutes since you placed the bar outside. You are given the following relationships: $l = 0.00013(t - 20) + 10$ and $n = 12t - 240$. Write a function f that gives the length of the bar, l, (in meters) in terms of the number of minutes elapsed, n, since the bar was placed outside.

64. Let $n = 240s + 123$ and let $t = 0.14s^2$. Write a function h that gives n in terms of t.

65. Let $160 = 2t + 4m$ and let $p = \frac{1}{4}tm^2$. Write a function k that gives p in terms of t.

VI. INVERSE FUNCTIONS: REVERSING THE PROCESS (TEXT: S6)

In Exercises #66-68, do the following.
     a. Describe the function process (for example, "the input is increased by 2 and then tripled").
     b. Describe the process that undoes the function (for example, if f is a process that increases its input by 7, the process that undoes f will decrease its input by 7).
     c. Algebraically define a process that undoes the process of the given function – that is, define the inverse function for each of the given functions.

66. $f(x) = \frac{7}{5}x$ 67. $g(x) = 10 + x$ 68. $h(x) = \frac{x}{12}$

69. a. Define algebraically a process f that multiplies its input by 3, and then decreases this result by 1.
    b. Describe the process that undoes the process of f described in part (a).
    c. Algebraically define the process f^{-1} that undoes the process of f that is described in part (a).

70. Given the function f defined by $f(x) = 3x + 2$,
    a. Define f^{-1}, the function that undoes the process of f.
    b. Show that $f\left(f^{-1}(x)\right) = x$ for the function f that is given above.
    c. Show that $f^{-1}\left(f(x)\right) = x$ for the function f that is given above.
    d. What do you conclude about the relationship between f and f^{-1}

71. Given $g(x) = 14x - 9$, what are the value(s) of the input x when $g(x) = 109$?

72. Given $h(x) = \frac{1}{6}x^2 + 7$, what are the value(s) of the input x when $h(x) = 31$?

73. Given that $h(x) = 4x$ determines the perimeter of a square given a square's side length x
    a. Algebraically define h^{-1} and describe its input quantity, output quantity, and process.
    b. Describe the input quantity, function process, and output quantity for $h\left(h^{-1}(x)\right)$.

74. Assume that g and z are functions that have inverse functions. Use the words *input* and *output* to express the meaning of the following expressions.
    a. $g^{-1}(21)$
    b. $z^{-1}(32) = 43$

In Exercises #75-78, find the inverse of each given function. Then, determine if the inverse that you found is a function. Justify your answer.

75. $g(x) = 2x + 4$ 76. $h(x) = 2x^3 - 6$ 77. $k(x) = \frac{1}{2}x^2 + 12$ 78. $m(x) = \frac{2}{x+3}$

79. After Julia had driven for half an hour, she was 155 miles from Denver. After driving 2 hours, she was 260 miles from Denver. Assume that Julia drove at a constant speed. Let f be a function that gives Julia's distance in miles from Denver after having driven for t hours.
    a. Determine a rule for the function f.
    b. Interpret the meaning of $f^{-1}(500)$ and then find its value.
    c. Determine a rule for f^{-1}.
    d. Construct a graph for $f^{-1}(d)$.

80. The functions f and g are defined in the given table. Based on this table, complete the following.

x	-3	-2	-1	0	1	2	3
$f(x)$	9	2	6	-4	-5	-8	-9
$g(x)$	3	0	3	2	-3	-1	-5

    a. Is f^{-1} a function? Explain why or why not.
    b. Is g^{-1} a function? Explain why or why not.
    c. Evaluate the following expressions:
        i. $f^{-1}(-9)$ ii. $g(3)$ iii. $f\left(g(2)\right)$ iv. $f^{-1}(2)$ v. $f^{-1}\left(g(3)\right)$ vi. $g\left(f^{-1}(-4)\right)$

79. The graphical representation of the functions g and h are given. Use this information to complete the following.
    a. As x increases from 0 to 5, how does $g(x)$ change? Be specific.
    b. As x increases from 0 to 4.5, how does $h(x)$ change?
    c. Use the graphs above to evaluate the following:
        i. $g^{-1}(2)$ ii. $g^{-1}\left(h(2)\right)$ iii. $g\left(h^{-1}(4)\right)$
    d. Is the inverse to h a function?

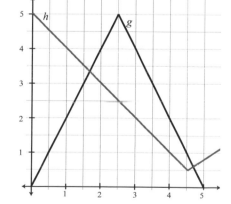

80. Given the functions $f(y) = \frac{3}{2}y + 21$ and $g(x) = 2x + 1$, answer the following questions.
    a. Determine f^{-1} and explain how f^{-1} relates to f in terms of the input values and output values of each function.
    b. Find $g^{-1}\left(g(2)\right)$. Find $g^{-1}\left(g(3.5)\right)$.
    c. What pattern do you observe about $g^{-1}\left(g(x)\right)$? Explain.   *Exercise continues on the next page.*

d. Determine the composite function $f(g(x))$ and explain, using input-output language, the meaning of the function $f(g(x))$ relative to the functions f and g.

e. Considering the description of the function $f(g(x))$ given in part (d), describe the inverse of this function. Justify whether $f^{-1}(g^{-1}(x))$ or $g^{-1}(f^{-1}(x))$ is the proper notation for representing this function's inverse, and then determine the rule of the chosen function.

f. What are the domain and range of the function f?

81. Apply the ideas of function inverse and function composition to answer each of the following.
   a. Define the area of a square A in terms of its perimeter p.
   b. Define the diameter of a circle d in terms of its circumference C.

82. Apply the ideas of function inverse and/or function composition to answer each of the following.
   a. Define the perimeter of a square p in terms of its area A.
   b. If the radius of a circle is growing at 8 cm per second, define the area of a circle A in terms of the amount of time t since the radius was 0 cm and started growing.
   c. Define the diameter of a circle d in terms of its area A.

VII. INTRODUCING THE DIFFERENCE QUOTIENT (TEXT: S7)

For Exercises #83-87 do the following.
   a. Determine the function's average rate of change from $x = 2$ to $x = 5$.
   b. Explain what your answer to part (a) represents.
   c. Write an expression that represents the function's average rate of change over any interval of length h. Simplify your expression if possible.

83. $f(x) = 12x + 6.5$ 84. $f(x) = 97$ 85. $f(x) = 6x^2 + 7x - 11$

86. $f(x) = 3x^3 - 9$ 87. $f(x) = \frac{1}{2x}$

88. Recall that the volume of a box V (measured in cubic inches) as a function of the length of the side of the square cutout x (measured in inches) is given by $f(x) = x(11 - 2x)(8.5 - 2x)$ with $V = f(x)$.
   a. Describe the meaning of each of the following expressions.

   i. $f(x + 3)$ ii. $f(x + 3) - f(x)$ iii. $\frac{f(x+3)-f(x)}{(x+3)-x}$

   b. Evaluate $\frac{f(x+3)-f(x)}{(x+3)-x}$ when $x = 0.5$. Describe the value's meaning in this context.

89. $f(x) = x^2 - 6x + 10$ represents the altitude of a US Air Force test plane (in thousands of feet) during a recent test flight, as a function of elapsed time (in minutes) since being released from its airborne launcher.
   a. Find the average rate of change of the plane's altitude with respect to time as the time varies from $x = 2$ minutes to $x = 2.1$ minutes. Show your work.
   b. What is the meaning of the *average rate of change* you determined in part (a) in this context?
   c. Explain the meaning of the expression $f(k + 0.1)$.
   d. Explain the meaning of the expression $f(k + 0.1) - f(k)$.
   e. Suppose you have been given the graph of $y = f(x)$ below

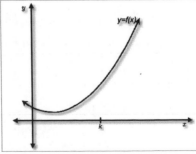

   and a specific value of x, say $x = k$. Show $f(k + 0.1) - f(k)$ on the given graph.

   f. What does the expression $\frac{f(k+0.1)-f(k)}{(k+0.1)-k}$ represent?

This investigation contains review and practice with important skills and procedures you may need in this module and future modules. Your instructor may assign this investigation as an introduction to the module or may ask you to complete select exercises "just in time" to help you when needed. Alternatively, you can complete these exercises on your own to help review important skills.

Meaning of Exponents
Use this section prior to the module or with/after Investigation 1.

Exponents are used to represent the number of identical factors in a product. For example, in the product $5 \cdot 5 \cdot 7 \cdot 7 \cdot 7$, the factor 5 appears twice and the factor 7 appears three times. Thus:

$$5 \cdot 5 \cdot 7 \cdot 7 \cdot 7 \text{ is equivalent to } 5^2 \cdot 7^3$$

Here are a few more examples. Note that each column represents the same value in three different forms (product notation, exponential notation, and decimal notation).

product notation	$4 \cdot 4 \cdot 4 \cdot 4 \cdot 4$	$2 \cdot 2 \cdot 2 \cdot 2 \cdot 5 \cdot 5 \cdot 5$	$7 \cdot 6 \cdot 4 \cdot 6 \cdot 7$
exponential notation	4^5	$2^4 \cdot 5^3$	$4 \cdot 6^2 \cdot 7^2$
decimal notation	1,024	2,000	7,056

In Exercises #1-3, rewrite each product in exponential and decimal notation.

1. $2 \cdot 6 \cdot 7 \cdot 7 \cdot 6 \cdot 6$ 2. $4 \cdot 5 \cdot 5 \cdot 4 \cdot 4 \cdot 4 \cdot 4$ 3. $5 \cdot 5 \cdot 5 \cdot 5 \cdot 5 \cdot 3 \cdot 7$

It's important to realize that exponents do <u>not</u> have to be whole numbers. It is possible to evaluate a^b where b is not a whole number. Use a calculator to evaluate each of the expressions in Exercises #4-6. Round your answers to three decimal places.

4. $8^{4.3}$ 5. $256^{0.5}$ 6. $80^{1.715}$

Even if you don't have a calculator, you should be able to estimate the value of exponential expressions with non-whole number exponents. For example, $2^{4.7}$ must be between 16 and 32 because $2^4 = 16$ and $2^5 = 32$. In Exercises #7-10 estimate the value of each exponential expression by giving two whole number values between which the value must fall.

7. $3^{3.4}$ 8. $10^{2.5}$ 9. $7^{0.8}$ 10. $4^{-1.9}$

The same number can be represented in terms of ANY base we choose. For example, consider the number 64. Each of the following exponential expressions represents a value of 64. *Take a moment to verify this.*

$$2^6 \quad 4^3 \quad 8^2 \quad 64^1 \quad 4{,}096^{1/2} \quad 16{,}777{,}216^{1/4}$$

However, the following expressions ALSO represent a value of 64 (up to a small rounding error). *Take a moment to verify this.*

$$3^{3.78558} \quad 5^{2.58406} \quad 10^{1.80618} \quad 15^{1.53575} \quad 28^{1.24809} \quad 100^{0.90309}$$

11. Which of the following exponential expressions also represent a value of 64 (up to a small rounding error)? Select all that apply.

     a. $7^{2.137}$ b. $11^{1.734}$ c. $9^{1.651}$ d. $\left(\frac{1}{3}\right)^{-3.786}$

12. Group the following exponential expressions based on their values (up to a small rounding error).

    a. $3^{3.72683}$ b. $12^{1.50415}$ c. $46^{1.02178}$ d. $15^{1.44459}$

    e. $10^{1.77815}$ f. $50^{0.95543}$ g. $80^{0.89274}$ h. $215^{0.76236}$

Properties of Exponents
Use this section prior to the module or with/after Investigation 2.

The following box shows the most common exponent properties we use to rewrite expressions.

Exponent Properties

For any real numbers a, b, m, and n, the following properties hold.

Product Rule: $a^m \cdot a^n = a^{m+n}$ **Quotient Rule:** $\dfrac{a^m}{a^n} = a^{m-n}$ (if $a \neq 0$)

Power Rule: $(a^m)^n = a^{m \cdot n}$ **Distributive Property:** $(ab)^m = a^m \cdot b^m$

Zero Exponent Rule: $a^0 = 1$. (if $a \neq 0$)   **Negative Exponent Rule:** $a^{-m} = \dfrac{1}{a^m}$ (if $a \neq 0$)

---

13. Write out each of the following in expanded form to demonstrate why the product rule $a^m \cdot a^n = a^{m+n}$ is true. *For example, $3^2 \cdot 3^4$ would be written as $3 \cdot 3 \cdot 3 \cdot 3 \cdot 3 \cdot 3$ in expanded form.*

    a. $2^3 \cdot 2^5$ b. $4^4 \cdot 4^3$ c. $7^6 \cdot 7^4$

14. Write out each of the following in expanded form to demonstrate why the product rule $(a^m)^n = a^{m \cdot n}$ is true.

    a. $(2^3)^2$ b. $(5^2)^3$ c. $(8^5)^2$

15. Write out each of the following in expanded form to demonstrate why the quotient rule $\dfrac{a^m}{a^n} = a^{m-n}$ is true.

    a. $\dfrac{5^6}{5^2}$ b. $\dfrac{3^7}{3^5}$ c. $\dfrac{8^6}{8^5}$

16. Use the quotient rule as well as the expanded form of the following expressions to demonstrate why it makes sense that $a^0 = 1$.

  a. $\dfrac{6^4}{6^4}$

  b. $\dfrac{2^5}{2^5}$

  c. $\dfrac{11^3}{11^3}$

In Exercises #17-28, use the exponent properties to rewrite each of the following using only positive exponents.

17. $\left(x^7\right)^3$

18. $\left(xy\right)^4\left(xy^2\right)$

19. $\left(r^2t\right)^3$

20. $\left(2h^3\right)\left(6h^2\right)$

21. $\left(x^3y^{-4}\right)\left(x^5y^2\right)$

22. $\left(p^2z^{-4}\right)^3$

23. $\left(a^4y\right)^2\left(a^5\right)^0$

24. $\left(4d^{-2}\right)\left(3n\right)^6$

25. $\dfrac{21x^{12}y}{x^7y}$

26. $\dfrac{14a^5p^2}{7ap^2}$

27. $\dfrac{x^4y^{10}z^3}{x^2y^5z^6}$

28. $\dfrac{9x^3y^4z^5}{12x^7y^3z^{11}}$

Solving Equations Involving Integer Exponents

Use this section prior to the module or with/after Investigation 4.

Consider an equation like $4x^3 + 9 = 35$. The solution is the value of x such that $f(x) = 4x^3 + 9$ outputs a value of 35.

$$4x^3 + 9 = 35$$
$$4x^3 = 26 \qquad \bullet \text{ subtract } 9$$
$$x^3 = 6.5 \qquad \bullet \text{ divide by } 4$$
$$x = \sqrt[3]{6.5}$$

The solution is $x = \sqrt[3]{6.5}$, or $x \approx 1.86626$.

In Exercises #29-32, solve each equation.

29. $2x^3 - 110 = 140$

30. $2(x^3 + 13) = -28$

31. $\frac{1}{2}x^5 - 20 = -1$

32. $3(x^5 - 8) + 10 = -182$

In Exercises #29-32 all of the exponents were odd integers. This is convenient because it created no doubt about the solution. If $x^3 = 64$, then $x = 4$ since $(4)^3 = 64$. If $x^3 = -64$, then $x = -4$ since $(-4)^3 = -64$. If the exponent is even this is not the case. For example, if $x^2 = 25$, then the solutions are $x = 5$ and $x = -5$ because $(5)^2 = 25$ AND $(-5)^2 = 25$. Therefore, if the exponent is even we have to consider that there are two possible solutions to the equation. For example, let's solve the equation $3x^4 - 10 = 7$.

$$3x^4 - 10 = 7$$
$$3x^4 = 17 \qquad \bullet \text{ add } 10$$
$$x^4 = \tfrac{17}{3} \qquad \bullet \text{ divide by } 3$$
$$x = \pm\sqrt[4]{\tfrac{17}{3}}$$

The solutions are both $x = \sqrt[4]{\tfrac{17}{3}}$ AND $x = -\sqrt[4]{\tfrac{17}{3}}$. Note that mathematicians often combine these two solutions into one statement by writing $x = \pm\sqrt[4]{\tfrac{17}{3}}$. It's important to recognize that "$\pm$" is indicating two different values for x.

In Exercises #33-36, solve each equation.

33. $\tfrac{1}{2}(x^2 + 5) = 82$ 34. $3x^2 - 13 = 41$ 35. $2x^4 + 25 = 187$ 36. $\tfrac{1}{4}x^6 - 5 = 5$

Solving Equations Involving Variable Exponents (without Logarithms)
Use this section prior to the module or with/after Investigation 5.

The previous two sections were based on solving equations involving power expressions (monomials). But sometimes we have equations where the exponent is unknown. For example, consider the equation $2 \cdot 3^x - 52 = 110$. Its solution is the value of x such that $2(3)^x - 52$ has a value of 110.

We start the solution process as shown.
$$2 \cdot 3^x - 52 = 110$$
$$2 \cdot 3^x = 162 \qquad \bullet \text{ add } 52$$
$$3^x = 81 \qquad \bullet \text{ divide by } 2$$

At this point we have no algebraic method for solving the equation. However, if you know your powers of 3, then you know (or can easily verify) that $3^4 = 81$. So the solution is $x = 4$.

If we instead ended up with something like $5^x = 112$, then for now we can estimate that the solution would be between $x = 2$ and $x = 3$ because $5^2 = 25$ and $5^3 = 125$.

In Exercises #37-40, solve each equation. If the solution is not an integer, then provide the two consecutive integers between which the answer must fall.

37. $3(2^x - 14) = 54$ 38. $\tfrac{1}{2}(3^x) + 17 = 52$ 39. $16(4^x) + 10 = 11$ 40. $6(3^x) - 2 = 28$

<table>
<tr><td align="center"><u>**Percentage**</u></td></tr>
<tr><td>A ***percentage*** refers to a type of measurement where the measurement unit is a specific proportion of a quantity. To be exact, **1% (of a quantity) is $\frac{1}{100}$ of the quantity's value.**</td></tr>
</table>

---

*1. After reading a scientific article on June 9th, 2020, Destiny told her friend Juan that the estimated world-wide death rate for COVID-19 cases up to that point was 5.65%.

a. What other information should Destiny have provided to make her statement more meaningful to Juan?

b. After Juan asked for clarification, Destiny conveyed that the estimated death rate for COVID-19 was 5.65% of the total number of positive COVID-19 cases since the pandemic began.

  i. What additional information did Destiny provide?

  ii. Does Juan have enough information to determine the number of COVID-19 deaths? Explain.

c. After learning that, as of June 9, 2020, there were a total of 7,246,000 people world-wide who had tested positive for COVID-19, Juan calculated the total number of deaths by multiplying 7,246,000 by 0.0565.

  i. Provide a rationale for rewriting 5.65% as 0.0565 prior to multiplying.

  ii. Explain why multiplying 7,246,000 by 0.0565 gives the number of COVID-19 deaths since June 9, 2020.

d. Discuss the meaning of percent and explain why it is important to include the "reference quantity" when reporting a percent. That is, explain why we must include the answer to the question, "Of what?" when stating a percentage ("percent (<u>of what?</u>)").

---

2. You likely encounter the idea of "percent (of something)" frequently in your daily life. For each of the following examples,

       (i)   perform the calculation,

      (ii) state the units on your answer, and

     (iii) use the definition of percent to explain why your approach makes sense.

   a. Determine the amount of a 15% tip on a food purchase of $85.

   b. Determine the number of hours left on your 9-hour phone battery that shows 27% of your battery usage remaining.

   c. Determine the number of people during the 2003 SARS pandemic who tested positive for SARS and survived, given that there were 8000 people who tested positive and 10% of the people who tested positive died.

   d. If you take out a 1-year loan for $6550 with interest charged at 23% of the amount borrowed, how much money must be repaid at the end of 1 year?

*3. The idea of percent (of something) allows us to make direct comparisons by considering the "parts per 100", or the number of $1/100^{ths}$ of one quantity compared to the number of $1/100^{ths}$ of another quantity.

   a. An acid mixture used by one lab technician contains 32.4 mL of acid combined with 90 mL of water. A second lab technician makes a mixture with 26.8 mL of acid and 67 mL of water. We want to know which technician made a "stronger" acid mixture.

     i. How can percentage calculations help us make the necessary comparison?

    ii. Which technician made the stronger acid mixture? Justify your answer.

b. The worldometer (a statistics reporting service) reported that the death rate for those testing positive for the 2003 SARS virus was 10% while the death rate of those testing positive for COVID-19 through June 9, 2020 was about 5%.

    i. Explain what this data tells us about the two virus outbreaks and what this data does NOT tell us about the outbreaks.

    ii. The 2003 SARS outbreak saw 8,000 people worldwide infected with the virus while, through June 9, 2020, experts calculated that there were 7,238,500 confirmed cases of COVID-19 worldwide. What else do we know about the two outbreaks based on the given information?

c. Come up with two examples of situations where you see percentages used in the real world and (i) explain what a percentage value represents in that context and (ii) the benefits of using a percentage measurement in that context.

*4. In this module we will often visualize the comparison of two measurements using the lengths of two line segments. This can achieve two things. First, it provides a way to check whether the results of a percentage calculation make sense. Second, it can help us think about a ***percent change*** in a quantity's value.

a. In order to make room for next year's models, a car dealership reduced the price of all new cars on their lot. One car that used to cost $24,995 now costs $22,355.

    new price: $22,355 _____

    old price: $24,995 _____

    i. How many times as large is the new price as the old price? How can you "see" this in the diagram?

    ii. *Fill in the blank*: "The new price is _____% of the old price." After filling in the blank, discuss with a partner or as a class how this question differs from part (a) and how we can "see" this value in the diagram?

b. Draw a vector on the diagram with a direction and length that represents the change in the car's price. What is the change in the car's price?

c. What is the change in the car's price as a percentage of the original price? [Note: This is called the *percent change* in the car's price.]

Percent Change

Percent change refers to the *difference* between two values of a quantity, measured as a percentage of the "starting" value.

For example, if the price of an item was $50 and increased to $59, then we can say the following.
- The change in price is $9.
- The percent change from the old price to the new price is $\frac{9}{50}$, or $\frac{18}{100}$, or 18% of the old price.
- The new price is $59, which is 118% of the old price of $50.

Note that it's common (and good practice) to describe a *decrease* as a negative change. For example, if the original price of an item was $80 and decreased to $62, then we can say the following.

- The change in price is –$18.

- The percent change from the original price to the new price is $-\frac{18}{80}$, or $-\frac{22.5}{100}$, or –22.5% of the original price.

- The new price is $62, which is 87.5% of the original price.

---

*5. You walk into a store that is having a sale. For each sale described, do the following. [*Draw diagrams to help you.*]
    (i) State the percent change in the price.
    (ii) State the number we can multiply the original price by to find the sale price.
    (iii) Find the sale price in dollars.

    a. original price: $150, sale: 40% off   b. original price: $915.99, sale: 15% off

*6. A store needs to raise prices on some of its items. For each price change described, do the following. [*Draw diagrams to help you.*]
    i) State the percent change in the price.
    ii) State the number we can multiply the original price by to find the new price.
    iii) Find the new price in dollars.

    a. original price: $52, increase: 10%   b. original price: $13, increase: 50%

---

7. In the early months of the COVID-19 outbreak in 2020 the Center for Disease Control (CDC) provided daily updates on the number of people testing positive for COVID-19 each day. Experts examined growth patterns in this data by comparing the number of new cases on a particular day with the number of new cases the prior day. The following table provides data for the state of New York for a 5-day period that began on March 15, 2020.

# of days since March 15, 2020 n	# of new cases c	Δc
0	975	
1	1717	
2	5400	
3	8452	
4	10445	
5	15885	

a. As a first step to examine this data, use the table to determine and record the daily change in the number of new cases, c, each day.

b. As a second step, determine the following:

i. On day 1 since March 15th, 2020, there were _____ times as many cases as there were on day 0. The percent change in cases from day 0 to day 1 was _____% (of _____).

ii. On day 2 since March 15th, 2020, there were _____ times as many cases as there were on day 1. The percent change in cases from day 1 to day 2 was _____% (of _____).

iii. On day 3 since March 15th, 2020, there were _____ times as many cases as there were on day 2. The percent change in cases from day 2 to day 3 was _____% (of _____).

iv. Explain what a negative percent change over a one day period conveys about the "growth" in number of cases over that one day period.

8. A new company expanded rapidly and doubled the number of its employees every month over the last year.

   a. Draw a diagram to represent what this means.

   b. *Fill in the blank*: The number of employees in one month was _____% of the number of employees in the previous month.

   c. What was the percent change in the number of employees from one month to the next? How can you represent this in your diagram?

9. A motion picture company released the new trailer for an upcoming movie yesterday. Today, its website received 5 times as many visitors as it did yesterday.

   a. Draw a diagram to represent what this means.

   b. What was the percent change in the number of visitors from yesterday to today?

   c. *Fill in the blank*: The number of visitors today was _____% of the number of visitors yesterday.

When examining situations with varying quantities, we typically want to identify pairs of quantities that change together (co-vary). When we identify such a pair, our next goal is to understand *how* their values change together and, if possible, to identify patterns in the relationship. One approach to studying patterns in co-varying relationships is to identify what stays the same as the values of the quantities change.

In a previous module we closely examined *linear functions*. A linear relationship exists when two quantities' values change together with a *constant rate of change*. In such a relationship the two quantities' values co-vary, but they must do so in such a way that the ratio of their changes $\frac{\Delta y}{\Delta x}$ is a constant. But not all co-varying relationships exhibit a constant rate of change. In this investigation we will explore patterns of change for a different type of relationship and identify what remains the same as two quantities co-vary.

For Exercises #1-5, use the following information. When a piece of new technology is invented and sold, its value decreases over time because it wears down and because even newer, better technology is developed to replace it. Suppose two electronic devices have the same resale value right now and that their resale values are expected to change according to the following patterns.

- **Product #1: iTech Device** The current resale value is $300. The resale value is expected to decrease 30% per year for the next several years.
- **Product #2: Dynasystems Device** The current resale value is $300. The resale value is expected to decrease $45 per year for the next several years.

*1. a. Find the expected resale value (in dollars) for both devices 1 year from now and describe how you determined each value.

b. Without performing additional calculations, predict which device will have a greater resale value 6 years from now.

c. Complete the given table showing the expected resale value for each device over the next 6 years.

Δt	Years from now, t	Expected resale value for the iTech Device, s (dollars)	Δs	Δt	Years from now, t	Expected resale value for the Dynasystems Device, n (dollars)	Δn
	0	300			0	300	
	1				1		
	2				2		
	3				3		
	4				4		
	5				5		
	6				6		

d. Check your prediction from part (b) by using the table in part (c).

---

e. The table on the left demonstrates how the expected resale value for the iTech Device co-varies with elapsed time. What stays the same as these two quantities' values change together?

f. The table on the right demonstrates how the expected resale value for the Dynasystems Device co-varies with elapsed time. What stays the same as these two quantities' values change together?

*2. a. Define function formulas to model the expected resale value in dollars for each device as functions of the number of years from now, t.

iTech Device: $s = f(t)$; where $f(t) =$ _____

Dynasystems Device: $n = g(t)$; where $g(t) =$ _____

b. Explain what each part of the formulas represents relative to the context we are exploring.

3. Find and interpret the following using your functions from Exercise #2.
   a. $f(4)$ b. $g(3)$

   c. $g(6) - g(2)$ d. $f^{-1}(147)$

*4. In Exercise #1, the table on the left has a column showing values of $s = f(t)$ (the expected resale value for the iTech Device in dollars) as well as a column showing values of Δs (the change in the expected resale value for the iTech Device in dollars).

a. What are the ratios of consecutive entries in the column for s? That is, what are the values of $\frac{f(1)}{f(0)}$, $\frac{f(2)}{f(1)}$, $\frac{f(3)}{f(2)}$, and so on?

b. How does your answer to part (a) relate to the original problem context? In other words, what do these ratios tell us about the resale value of the iTech Device?

c. What are the ratios of values of Δs compared to the values of s at the beginning of each interval?

d. How does your answer to part (c) relate to the original problem context? In other words, what do these ratios tell us about the resale value of the iTech Device?

The following are graphs of functions f and g that model the expected resale values for each device over the next 6 years. On each graph we have plotted the points that correspond to the ordered pairs you should have calculated in Exercise #1.

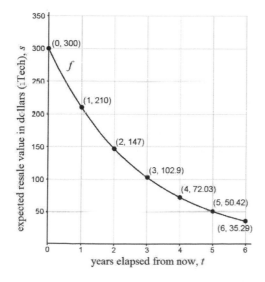

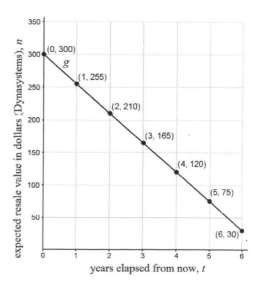

To help focus attention on the values of s and n (the expected resale values of each device in dollars) represented in our previous tables, we have drawn vertical line segments whose lengths represent values of s and n respectively.

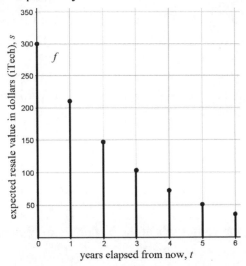 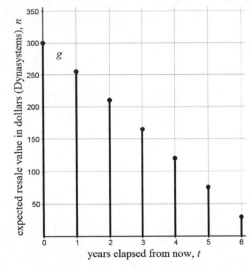

*5. a. Explain how we can "see" on the above graph on the left the fact that the expected resale value of the iTech Device at some moment is 70% of its expected resale value one year earlier.

   b. Explain how we can "see" on the above graph on the left the fact that the expected resale value of the iTech Device changes by –30% each year.

   c. Explain how we can "see" on the above graphs that only one of the two devices has an expected resale value that changes at a constant rate of change with respect to the number of elapsed years.

6. Suppose another piece of technology (the iTech2 Device) has a current resale value of $400, and that this value is expected to change by –50% per year. Let function h model its expected resale value in dollars t years from now.

   a. *Without performing any calculations*, plot the approximate values for the ordered pairs $(1, h(1))$, $(2, h(2))$, $(3, h(3))$, $(4, h(4))$, $(5, h(5))$, and $(6, h(6))$.

   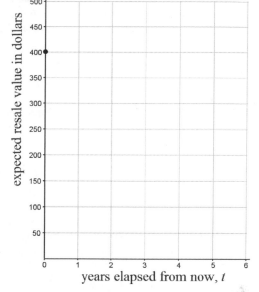

   b. Explain how you were able to estimate the values for the ordered pairs in part (a).

   c. Suppose that another device (TecTone) has a current resale value of $500 and its resale value is expected to decrease by 20% each year. Let function j model its expected resale value t years from now. Plot the point (0, 500) on the graph in part (a), then *without performing any calculations*, plot the approximate values for the ordered pairs $(1, j(1))$, $(2, j(2))$, $(3, j(3))$, $(4, j(4))$, $(5, j(5))$, and $(6, j(6))$.

   d. Write formulas to represent functions h and j.

Exponential Functions

When two quantities change together such that, for equal changes in one quantity's values, the second quantity's values have a ***constant percent change***, then we say that the relationship is an ***exponential function***.

---

*7 In Exercises #1-6 you explored several examples of relationships that can be modeled with an exponential function. Using your work with these examples, explain why exponential functions ***do not*** have a constant rate of change of one quantity with respect to the other quantity.

*8. From about 2000 to 2006, home prices in the United States increased dramatically prior to the real estate "crash" that began around 2007-2008. Suppose that from 2000 to 2006 the value of a specific house in California increased by 25% per year.

    a. Without performing any calculations, predict the home's value after 6 years if its value increases 25% per year. (Will the value be 50% larger? Double the original value? Something else?)

    b. Complete the following table and draw a graph modeling the value of this house in dollars *n* years since the beginning of 2000. Be sure to label the axes.

years since the beginning of 2000	home price (in dollars)
0	210,000
1	
2	
3	
4	
5	
6	

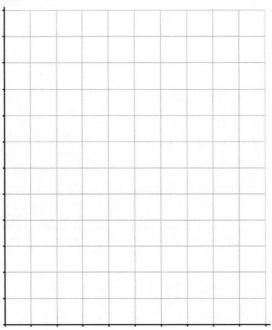

    c. By what number can we multiply the value of the home at one moment in time to find its value one year later?

    d. Fill in the blank: The value of the home at one moment in time is _____% of its value one year earlier.

    e. Explain how we can "see" the value of the multiplier you identified in part (c) in the graph from part (b).

    f. The home's value at the beginning of 2006 was how many times as large as its value at the beginning of 2000? How does this compare to your prediction in part (a)?

In the previous investigations we established the meaning
of exponential growth. In this investigation we will
explore the implications of exponential growth in various
situations and formalize the idea of a ***growth factor***.

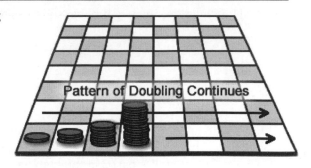

Pattern of Doubling Continues

*1. How large do numbers become if you repeatedly
     double them? Imagine you have a chessboard and
     place 2 pennies on the first square and double that to
     4 pennies on the second square, double that to 8
     pennies on the third square, and so on.

   a. There are 64 squares on a chessboard. <u>Without doing any calculations</u>, make two <u>predictions</u>.
       i. About how many pennies do you think will be on the last square? Will it match something like
           the number of students at your university? The population of Los Angeles? The population of
           the United States? Something else?

       ii. How tall will the stack of pennies be on the 64th square? As tall as a house? As tall as a
           skyscraper? The height of Mount Everest? Something else?

   b. Fill in the given table and use it to create a formula and graph relating the number of pennies on a
       square with the square's position number. Be sure to label the axes.
       <u>Table:</u>   <u>Graph:</u>   <u>Formula (define
                                                                                           your variables):</u>

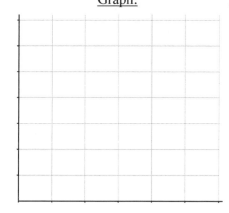

Square Number	Number of Pennies
1	
2	
3	
4	
…	
10	
…	
30	

   c. How do the number of pennies on the chessboard vary as you move along the board? For
       example,
       i. each time the square position increases by 1, what happens to the number of pennies?

       ii. each time the square position increases by 2, what happens to the number of pennies?

       iii. each time the square position increases by 3, what happens to the number of pennies?

   d. What is the *<u>percent change</u>* in the number of pennies on a square when the square position
       increases by 1?

e. How many pennies will be on square 64? How does this compare to your estimate in part (a)?

2. The function $f(x) = 2^x$ is an example of an exponential function, and some people call this a *doubling function*. The function in Exercise #1 was a doubling function, but its domain was restricted to positive integers. We can also think about continuous doubling functions.

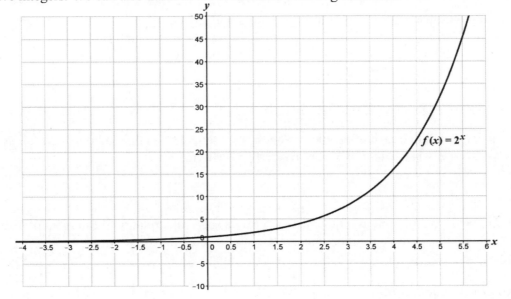

a. Determine the value of $f(0)$.

b. i. When x increases by 1, the new output value is _____% of the previous output value.

    ii. What is the percent change in the output value when x increases by 1?

c. As x increases away from $x = 0$, how do the function values change? What happens as x gets very, very large?

d. As x decreases away from $x = 0$, how do the function values change? What happens as x gets very, very small (becomes a large magnitude negative number)?

*3. View the Module 4 Investigation 3 PowerPoint animation titled **Exercise 3 Animation**.
   a. What idea or ideas are being demonstrated in this animation?

   b. How is this image different from thinking about the function using a table of values like the one shown?

x	0	1	2	3	4	5
$f(x)$	1	2	4	8	16	32

4. Let $g(x) = 3^x$ and $h(x) = 6\left(\frac{2}{3}\right)^x$.

   a. Add these functions to the graph in Exercise #2.

   b. With a partner or as a class, discuss how these functions compare to each other and to $f(x) = 2^x$. In particular, discuss the following questions.

      i. What is the vertical intercept for each function?

      ii. As x increases, how do the function values change? What happens as x gets very large?

      iii. As x decreases, how do the function values change? What happens as x gets very small (becomes a large magnitude negative number)?

      iv. What is the 1-unit percent change in the function values?

(1-Unit) Growth Factor

For an exponential function, the ratio of output values is always the same for equal-sized intervals of the domain. When x increases from x_1 to x_2, this ratio $\frac{f(x_2)}{f(x_1)}$ is called a ***growth factor***. If the interval is 1-unit wide (that is, if the change in x from x_1 to x_2 is 1), then the ratio is the ***1-unit growth factor***.

The growth factor is a useful number because it tells us how many times as large $f(x_2)$ is compared to $f(x_1)$, and thus can be used to determine $f(x_2)$ if we know $f(x_1)$.

Note that if this factor is between 0 and 1 it is often called a ***decay factor*** because a multiplier between 0 and 1 means that the function values are decreasing as x increases.

---

*5. The total number of COVID-19 cases in New York state increased by 76% each day between March 16, 2020 and March 20, 2020. There were 975 cases on March 16, 2020.

a. By what factor do you multiply 975 to determine the number of cases on March 17, 2020?

b. How many total cases were in New York state on March 20, 2020?

c. Suppose this trend continued past March 20.
   i. Define a function f to model the total number of cases of COVID-19 in terms of the number of days since March 16, 2020. Be sure to define your variables.

   ii. Evaluate $f(7)$ and explain what it represents.

   iii. The population of New York state at the time was about 19,450,000. If the trend continued according to this model, how long would it have taken for all of the people in New York to become infected? [*First make a prediction and then use calculations or a graphing calculator to determine the answer.*]

It's worth noting that patterns can change due to interventions (such as social distancing) or other factors (such as infected people increasingly coming into contact with more infected people and fewer unaffected individuals). Therefore, in mathematical modeling, it's important to continue to reexamine patterns of change as new data is available.

*6. Some of your classmates made the following claim. "The functions $f(x) = x^2$ and $g(x) = x^3$ are some other examples of exponential functions."
a. Are your classmates correct? Justify your answer using what you've learned so far in this module.

b. If you agree with your classmates, then come up with at least two more examples of exponential functions. If you disagree with your classmates, then give a possible reason for why this might be a common mistake for students.

In Exercises #7-10, do the following.
    a. Find the ratio of output values that correspond to increases of 1 in the input value (this value is the 1-unit growth/decay factor).
    b. Determine the 1-unit percent change by comparing the change in the output values to the function value at the beginning of a 1-unit interval for x.
    c. Identify or determine the value of the function when $x = 0$.
    d. Use the information from parts (a) through (c) to define a function formula for the relationship.

*7.

x	0	1	2	3
$f(x)$	16	4	1	0.25

    a.

    b.

    c.

    d.

8.

x	1	2	3	8
$g(x)$	260	299	343.85	691.605

    a.

    b.

    c.

    d.

9.

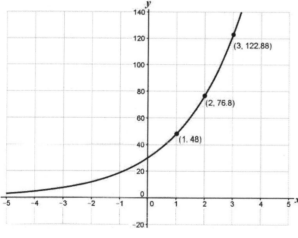

    a.

    b.

    c.

    d.

*10.

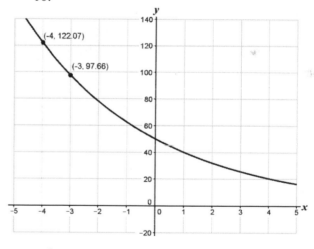

    a.

    b.

    c.

    d.

11. Let $f(x) = 34(1.19)^x$.

   a. What does the number "34" represent for this function?

   b. What does the number "1.19" represent for this function?

   c. Fill in the blank: Whenever x increases by 1, the new output value is _____% of the old output value.

   d. What is the 1-unit percent change and what does it tell us?

*12. Let $g(p) = 1.578(0.68)^p$.

   a. What does the number "1.578" represent for this function?

   b. What does the number "0.68" represent for this function?

   c. *Fill in the blank*: Whenever p increases by 1, the new output value is _____% of the old output value.

   d. What is the 1-unit percent change and what does it tell us?

13. After having 1.4 million people at the start of 2010, the population of a city has been decreasing by 2.1% per year.

   a. What is the 1-year percent change in the city's population?

   b. *Fill in the blank*: When the time elapsed since the beginning of 2010 increases by 1 year, the new population is _____% of the old population.

   c. What is the 1-year growth or decay factor and what does this value tell us about the situation?

   d. Write a function formula to model the city's population (in millions) in terms of the time elapsed since the beginning of 2010 (in years).

*1. Express your understanding of exponents by answering the following questions.
   a. i. In the term b^x what does x represent?

   ii. What does b^5 represent?

   iii. Evaluate 4^3 and say what your answer represents.

   b. Calculate the following: i. $9^{1/2}$ ii. $4^{1/2}$ iii. $121^{1/2}$

   c. What do your answers in (b) represent? In general, how do you go about determining the value of some number b raised to the ½ power?

   d. In general, how do you go about determining the value of $x^{1/3}$?

   e. Calculate $4^{3/2}$ (also represented as $4^{1.5}$). What does your answer represent?

   f. Solve the following equations for x:
      i. $x^2 = 81$ ii. $x^3 = 27$ iii. $x^{3/4} = 8$ iv. $x^{1/5} = 2$

   g. Simplify: $a^{2.5} \cdot a^4$. Describe what your answer represents.

   h. Simplify: $\dfrac{b^{1.8}}{b^2}$. Describe what your answer represents.

   i. Simplify: $(c^2)^{3.1}$. Describe what your answer represents.

*2. You are given that (4, 38) and (7, 50) are two ordered pairs for an *exponential growth function*. Your classmate noticed that the difference in the outputs is 12 as the inputs change by 3, and created the table showing some additional ordered pairs.

x	y
4	38
5	**42**
6	**46**
7	50

Explain how you know that the two additional ordered pairs (bolded) cannot be accurate even though we don't have the function formula or the graph for this relationship.

3. From March 16, 2020 to March 20, 2020, the total number of cases of COVID-19 tripled every 2 days in New York state, following a pattern of exponential growth.
   a. What was the 1-day growth factor and 1-day percent change during this period for the total number of COVID-19 cases in New York state?

   b. What was the 4-day growth factor and 4-day percent change during this period for the total number of COVID-19 cases in New York state?

   c. There were 975 total cases of COVID-19 on March 16, 2020 in New York state. Assuming the pattern of growth continued, give two methods (based on your work in parts (a) and (b)) to determine the total number of cases in New York state on March 24, 2020.

*4. When the COVID-19 pandemic spread across the United States, many states implemented social distancing measures and stay at home orders to slow its spread. After a few weeks, however, many states began to reopen businesses and relax these measures. In many states this caused an increase in the infection rate.

   After relaxing stay at home orders and social distancing measures, the number of total COVID-19 cases in Arizona increased from 16,891 on May 23, 2020 to 31,264 on June 11, 2020 [a period of 20 days]. *For this exercise, assume that the growth in the total cases followed an exponential model.*

   a. If this growth pattern continued, how many total cases would Arizona have had 20 days after June 11, 2020?

   b. By what percent did the total number of Arizona COVID-19 cases increase during this 20-day period?

   c. Assuming an exponential growth model, what was the 1-day growth factor and 1-day percent change for the total number of COVID-19 cases in Arizona during this period?

   d. Assuming an exponential growth model, what was the 10-day growth factor and 10-day percent change for the total number of COVID-19 cases in Arizona during this period? What about the 30-day growth factor and 30-day percent change? [*Try to think about two different ways to determine these values.*]

   10-day growth factor: _____   30-day growth factor: _____

   10-day percent change: _____   30-day percent change: _____

e. Define a function *f* that expresses the total number of Arizona COVID-19 cases during this period in terms of the number of days since May 23, 2020.

f. Use your function to determine each of the following values and explain what they represent.
   i. *f*(4)   ii. *f*(13)   iii. *f*(26)

*5. A farm in Canada had 218 alpacas on January 1, 2004. After 8 years the alpaca population decreased to 187. Assume the number of alpacas decays exponentially.
   a. What is the 8-year growth/decay factor?

   b. Fill in the blank: The number of alpaca on the farm on January 1, 2012 was _____% of the number of alpaca on the farm on January 1, 2004.

   c. What is the 8-year percent change?

   d. Assuming the alpaca population continues to be modeled by the same exponential model, how many alpacas can we expect to be on the farm on January 1, 2020?

   e. Since 8-year changes in time are quite long, we might want to know how the population of alpacas changes over shorter periods of time. What is the 1-year growth/decay factor?

   f. What is the 1-year percent change?

   g. Define a function *f* that relates the number of alpacas on the Canadian farm *t* years from January 1, 2004 (Assume the alpaca population continues to change by the same decay factor each year).

*6. Assume the number of alpacas continues to change by the same decay factor each year as defined in Exercise #5. Use your calculator determine the following;
   a. After how many years will there be 109 alpacas remaining on the farm, assuming the herd started with 218 alpacas?

   b. After how many total years will there be 55 alpacas?

   c. After how many years will a herd of 120 alpacas decrease to 60 alpacas?

*7. Assume the number of alpacas continues to change by the same decay factor each year as defined in Exercise #5.

    a.   What is the 1-month growth/decay factor?

    b.   Define a function g that relates the number of alpacas on the Canadian farm k months from January 1, 2004 (Assume the alpaca population continues to change by the same decay factor each year).

    c.   What is the 2-month growth/decay factor?

8. After taking medicine, your body begins to break it down and remove it according to a pattern of exponential decay. Suppose you take 500 mg of Ibuprofen and, after 5 hours, only 110 mg is left in your system.

    a.   After 5 hours, what percent of the 500 mg dose remains in your body? How much Ibuprofen (in mg) remains in your body?

    b.   Define a function f that expresses the amount of Ibuprofen in milligrams (mg), B, present after n 5-hour time intervals since taking the medicine.

    c.   Define a function g that expresses the amount of Ibuprofen in milligrams (mg), B, present t hours after taking the medicine.

    d.   How much Ibuprofen remains in your body 3 hours after taking the medicine? After 12 hours?

    e.   How does the *change* in the amount of Ibuprofen remaining change as the number of hours elapsed increases?

For Exercises #1-2, determine the specified growth or decay factor, percent change, and initial value for each of the following exponential functions.

*1. $f(x) = 9.5(1.24)^x$

    a. 1/2-unit Growth Factor:

    b. 2-unit Growth Factor:

    c. 2-unit Percent Change:

    d. Initial Value:

*2. $g(x) = 0.46(0.874)^{4x}$

    a. 1/4-unit Decay Factor:

    b. 1-unit Decay Factor:

    c. 5-unit percent change:

    d. Initial Value:

In Exercises #3-4, determine the specified growth or decay factor, percent change, initial value, and function formula for each of the tables modeled by exponential functions.

3.

x	0	2	4	6
$f(x)$	16	10.24	6.554	4.194

    a. 2-unit Decay Factor:

    b. 1-unit Percent Change:

    c. 3-unit Decay Factor:

    d. ½ -unit Decay Factor:

    e. Initial Value:

    f. Formula:

*4.

x	1	4	7	10
$g(x)$	260	278.2	297.674	318.511

    a. 3-unit Growth Factor:

    b. 6-unit Percent Change:

    c. 1-unit Growth Factor:

    d. ¼-unit Growth Factor:

    e. Initial Value:

    f. Formula:

In Exercises #5-6, determine the specified growth or decay factor, percent change, initial value, average rate of change, and function formula for the exponential function with the graph.

*5. Use the graph to complete the following.
    a. 2-unit Growth/Decay Factor:

    b. 6-unit Growth/Decay Factor:

    c. 1-unit Percent Change:

    d. ½-unit Growth/Decay Factor:

    e. 3-unit Growth/Decay Factor:

    f. Initial Value:

    g. Formula:

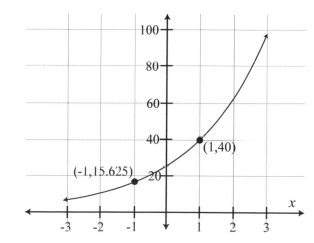

    h. The average rate of change of $r(x)$ with respect to x as the value of x increases from -1 to 1.

    i. Explain how to interpret the value you found in (h).

*6. Use the graph to complete the following.
   a.   3-unit Growth/Decay Factor:

   b.   6-unit Growth/Decay Factor:

   c.   1-unit Percent Change:

   d.   ½-unit Growth/Decay Factor:

   e.   5-unit Growth/Decay Factor:

   f.   Initial Value:

   g.   Formula:

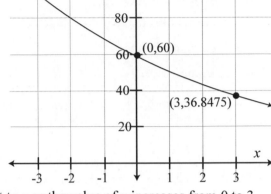

   h.   The average rate of change of $w(x)$ with respect to x as the value of x increases from 0 to 3.

   i.   Explain how to interpret the value you found in (h).

In Exercises #7-9, determine the growth or decay factor, percent change, initial value and function formula for each exponential situation.

*7. For the years 2006 to 2010 (which includes **five** years of data), the number of people in the USA living in poverty increased exponentially from 36.46 million people to 46.34 million people.
   a. 5-year Growth Factor: b. 1-year Percent Change:

   c. Initial Value: d. Function:

8. After having 1.97 million people in 2010, the population of a city has been decreasing by 4.2% every three years.
   a. 1-year Decay Factor: b. ¼ -year Percent Change:

   c. Initial Value: d. Function:

*9. The amount of caffeine in your body decreases by 27% every 4 hours.
   a.   24-hour Decay Factor: b.   24-hour Percent Change:

   c.   1-hour Decay Factor: d.   1-hour Percent Change:

10. Consider an exponential relationship in which b represents the 1-unit growth factor, c represents the n-unit growth factor, and d represents the m-unit growth factor.
   a.   Write a formula that expresses the n-unit growth factor in terms of the 1-unit growth factor.

   b.   Write a formula that expresses the 1-unit growth factor in terms of an n-unit growth factor.

   c.   Write a formula that expresses the m-unit growth factor in terms of an n-unit growth factor.

When you deposit money in a bank they pay you interest based on the balance. The interest can be paid in many different ways such as once a year, twice a year, once a month, etc. By convention, banks calculate the interest rate per compounding period by dividing the advertised annual interest rate (often called the Annual Percentage Rate or APR) by the number of compounding periods in one year. *Note that this is a different technique from determining partial growth factors.*

*1. Suppose you are going to invest $1000. You have three choices on how to invest your money and in all three situations the annual interest rate (APR) is 8%.

    a. The first account advertises, "8% APR compounded annually". This means that at the end of each year 8% of the current balance is added to the value of the account.

       i. What is the interest rate per compounding period?

       ii. Complete the following table.

8% Compounded Annually	
number of years since investment was made	value of the investment (in dollars)
0	
1	
2	
t	

       iii. What is the annual growth factor for this account? Interpret the annual growth factor in the context of this problem.

    b. The second account advertises "8% APR compounded semiannually". This means that interest is added two times per year (every six months).

       i. What is the interest rate per compounding period?

       ii. Complete the following table.

8% Compounded Semiannually	
number of years since investment was made	value of the investment (in dollars)
0	
0.5	
1	
2	
t	

       iii. What is the six-month growth factor for this account?

iv. What is the annual growth factor for this account? Interpret the annual growth factor in the context of this problem.

v. Use the annual growth factor computed in part (iv) to determine the actual percentage change of the account over a 1-year time period. This percentage is often referenced as the annual percentage yield (APY).

c. The third account advertises "8% APR compounded daily". This means that interest is added 365 times per year (every day).
   i. What is the interest rate per compounding period?

   ii. Complete the following table.

8% Compounded Daily	
number of years since investment was made	value of the investment (in dollars)
0	
$\frac{1}{365}$	
$\frac{2}{365}$	
1	
2	
t	

   iii. What is the daily growth factor for this account?

   iv. What is the annual growth factor for this account? Interpret the annual growth factor in the context of this problem.

   v. Use the annual growth factor computed in part (iv) to determine the annual percentage yield (APY).

d. Which of the three accounts would you chose for your investment? Justify your answer.

*2. A bank offers an annual interest rate of r compounded n times per year, where r is the APR expressed as a decimal. Let a be the initial value of the investment. Define a function g that determines the value of the investment B in terms of the number of years since the investment was made, t.

3. Bank USA is offering a 9% annual rate compounded monthly and Southwest Investment Bank is offering an annual rate of 8.5% compounded daily. Which bank would you choose for your investment?

*4. You should have noticed that the techniques in this investigation for finding growth factors related to different units of time are not the same as the techniques used in previous investigations. This is because bank policy differs from our goals in creating partial growth factors. Look at the following comparison.

<table>
<tr><th>Non-Financial Contexts</th><th>Compound Interest</th></tr>
<tr><td>

The population of a certain country is 26.4 million people and is increasing by 4% per year.

Annual growth factor: 1.04

Annual percent change: 4%

Monthly growth factor:
$$(1.04)^{1/12} \approx 1.003274$$

Monthly percent change: 0.3274%

When the monthly growth factor is applied 12 times it produces the annual growth factor.
$$[(1.04)^{1/12}]^{12} = 1.04$$

</td><td>

$1200 is deposited into an account with an APR of 4%.

If interest is compounded once per year, the annual growth factor is: 1.04

The annual percent change is: 4%

If interest is compounded once per month, the monthly growth factor is: $1 + \frac{0.04}{12} = 1.00\overline{3}$

The monthly percent change is: $0.\overline{3}\%$

If interest is compounded once per month, and we do this for one year, the annual growth factor is different than when interest is compounded only once per year.
$$(1.00\overline{3})^{12} \approx 1.04074$$

</td></tr>
</table>

a. In non-financial contexts, we know the actual annual growth factor based on data and we cannot change the long-term growth pattern of the function relationship. How does this play a role in the methods we use for computing growth factors for different-sized changes in the input quantity's value?

b. Why don't banks have the same restrictions as those described in part (a) and how might this impact their techniques?

*5. A deposit of $5,000 is made into an account paying an APR of 5%. Determine the amount in the account 10 years after the investment was made if the interest is compounded:
   a. Annually

   b. Monthly

   c. Weekly

   d. Daily

*1. $1000 is invested into an account with an annual percent rate (APR) of 8%. Given the compounding period, determine the account value at the end of the fifth year. Also determine the annual percent yield (APY) for the given compounding period. (*Round to at least 3 decimal digits.*)

	APY	Difference in the value of the APY	Value of Investment after 5 years	Difference in the value of the investment
Compounded Yearly				
Compounded Quarterly				
Compounded Monthly				
Compounded Daily				
Compounded Hourly				
Compounded Every Minute				

2. How is the number of compounding periods per year related to the annual growth factor and the APY? Why do you think this is happening?

*3. a. What do you anticipate the APY will be if the APR is 8% and the investment is compounded every second? Every 1/100th of a second?

   b. What do you predict is the largest annual growth factor possible given an APR of 8%? Interpret the meaning of this value in the context of this situation.

*4. Consider investment APRs of 5%, 6%, 7%, and 8%.
   a. Complete the following table by predicting the largest annual growth factor that corresponds to those APRs. (Round to at least 6 decimal digits)

APR	5%	6%	7%	8%
Largest Annual Growth Factor, g				
$g^{1/r}$ (where r is the APR as a decimal)				

   b. What do you notice about the values in the second row of the table?

An interesting mathematical idea is that no matter what the APR is for a given investment, the largest annual growth factor for that investment is related to the number 2.71828… This irrational number is referred to as *e*.

*5. The value of *e* is equal to $g^{1/r}$ where *g* is the largest annual growth factor and *r* is the decimal value of the given APR. Undo this process to find an equation for *g* in terms of *e* and *r*.

*6. An investment of $1000 was made in four different banks at the given APRs. Complete the table by finding the largest annual growth factor for each APR. This value should be a decimal. Since this decimal is a rounded value (and therefore not exact), complete the second row by writing an expression that determines the exact value of the largest annual growth factor. Finally, write a formula that can be used to find the maximum value of the investment after *t* years (the value of the investment assuming the largest annual growth factor).

APR	5%	6%	7%	8%
Largest Annual Growth Factor (decimal form)				
Exact Representation of the Largest Annual Growth Factor				
Maximum Value of Investment After *t* Years				

*7. Define a function *f* that gives the value of an investment, *B*, in terms of the number of years since the initial investment was made, *t*. Let *a* represent the value of the initial investment and let *r* represent the decimal form of the APR.

Another way to think about the largest annual growth factor is to say that it is the growth factor that corresponds to compounding as many times as possible in a year. In other words, if the investment were to be compounded more often than daily, more often than hourly, more often than every second, etc., then we would say the investment is being ***compounded continuously***.

8. $1500 is invested into an account with an advertised APR of 7% compounded continuously.
   a. Determine the value of the investment after 6 years.

   b. Determine the amount of time required for the value of the investment to double.

   c. Determine the value of the account after 6 years had the investment been compounded quarterly instead of continuously.

Up until this point we have been unable to algebraically solve equations where the unknown is in the exponent. We utilized our graphing calculators in order to solve these equations. In this investigation we will learn how to solve these equations algebraically.

*1. a. Without a calculator approximate the solution to the following equations. (*Think about what value of x makes each equation true.*)

   i. $2^x = 10$ ii. $17^x = 10$

   b. Describe the process (thinking) you used to determine your solutions in part (a).

   c. What does each of your solutions in part (a) represent?

   d. For the equations in part (a), what information was needed to determine (or estimate) the value of the exponent.

Determining the solutions of the equations in Exercise #1 involved undoing the process of exponentiation. Instead of raising a base to an exponent we determined the exponent to which a number (the base) is raised to obtain some number.

As seen in Exercise #1, it can be difficult to determine the exact number of times a base value is a factor of some other number. The standard way of expressing the specific value of an unknown exponent, when provided the base value and the result of exponentiation, utilizes what we call *logarithmic notation*. As an example, to represent the number of times 5 is a factor of 125, we write $\log_5(125)$. In general, we say, $\log_b(m)$ represents the number of times b is a factor of m.

   e. Using logarithmic notation, represent the exact solutions to the equations in part (a).

   f. Describe what each of the following logarithmic expressions represents.

   i. $\log_{16}(94)$ ii. $\log_{3.4}(17.2)$ iii. $\log_{19}(2.7)$

The function that undoes the process of exponentiation is called the *logarithmic function*. The input of the logarithmic function is the output of the exponential function (the result of raising some base to an exponent) and the output of the logarithmic function is the input of the exponential function (the exponent to which the base is raised). We use "log" as an abbreviation for "logarithm" in expressions.

As an example, $\log_2(10) = x$ is read "log base 2 of 10 equals x" and is equivalent to $2^x = 10$ in exponential form. The number 10 represents the result of raising 2 to an exponent x. The value of x can be determined by considering the exponent to which 2 is raised to obtain the value 10.

2. Rewrite each of these equations in logarithmic form (if possible). If it is not possible, say why.
   a. $4^x = 64$
   b. $5^x = \frac{1}{125}$
   c. $2^x = -32$

We can rewrite any exponential equation in logarithmic form. *Note that the input to the logarithmic function can only be non-negative real numbers.*

Logarithmic Function

For $x > 0$ and $b > 0$, with $b \neq 1$, $y = \log_b x$ is equivalent to $b^y = x$. The function $f(x) = \log_b(x)$ is the logarithmic function with base b.

Note that logarithmic functions are the inverses of exponential functions with the same base.

---

*3. Rewrite the following exponential equations in logarithmic form.
   a. $y = 4^x$
   b. $b = 1.5(5)^a$
   c. $m = 2^{4t}$
   d. $q = 3(5)^{2k}$

*4. Without using your calculator, determine/estimate the value of the variable that makes the equation true.
   a. $\log_2 4 = y$
   b. $\log_9 \left(\frac{1}{81}\right) = t$
   c. $\log_3(-2) = k$
   d. $\log_5 10 = s$

*5. Logarithmic functions are the inverses of exponential functions. That is, both functions show the same relationship between two quantities, but the input and output quantities are switched.
   a. The graph of $f(x) = \log_4(x)$ is given.
      i. Plot the points $(x, f(x))$ when $x = \frac{1}{2}$, 1, 4, 16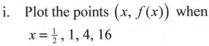

      ii. Explain how you know that f is an increasing function.

      iii. Explain how you know that y increases less and less, for equal increases in x.

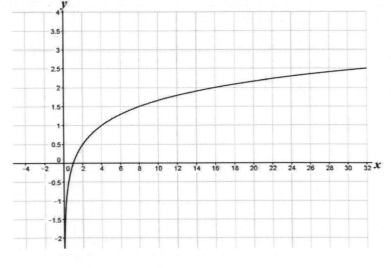

   b. Determine the average rate of change of y with respect to x as:
      i. x increases from 2 to 4
      ii. x increases from 4 to 6

---

c.   On the axes in part (a), draw the graph of $g(x) = \log_2(x)$.

d.   On the axes in part (a), draw the graph of $h(x) = \log_{0.5}(x)$.

Recall that in Module 4 Investigation 7 we defined the irrational number $e = 2.718281828\ldots$ A logarithm with base e or $\log_e(c) = x$ is commonly called the **natural logarithm** and is abbreviated "ln".

*Note: Your calculator has two logarithm buttons, log and ln. Even though we can use any base (greater than zero) in a logarithmic function, some calculators only evaluate logarithms for two bases: base 10 and base e. The log x button evaluates $f(x) = \log_{10}(x)$ and is referred to as the **common log**. The ln x button evaluates $\log_e(x)$ and is referred to as the **natural log** of x.*

6.   Convert the following to exponential form and evaluate/estimate the value of the unknown. Check your answer using your calculator.

a. $\log\left(\frac{1}{100}\right) = x$ b. $\ln(e^2) = k$ c. $\ln(7) = t$ d. $\log(1000) = s$

Recall the Properties of Exponents:

- **Property of Exponents #1:** $b^x \cdot b^y = b^{x+y}$

- **Property of Exponents #2:** $\dfrac{b^x}{b^y} = b^{x-y}$ for $b \neq 0$.

- **Property of Exponents #3:** $\left(b^x\right)^y = b^{xy}$

*7. Use your understanding of exponents and the meaning of b^x to justify each property stated above.

Because the result from evaluating a logarithmic expression (and the output of a logarithmic function) represents an exponent, the properties of exponents also apply to logarithms, with the rules of exponents expressed using logarithms and logarithmic form.

Logarithm Properties

Property #1: $\log_b(m \cdot n) = \log_b(m) + \log_b(n)$ for any $b > 0$, $b \neq 1$, and $m, n > 0$

Property #2: $\log_b\left(\dfrac{m}{n}\right) = \log_b(m) - \log_b(n)$ for any $b > 0$, $b \neq 1$, and $m, n > 0$

Property #3: $\log_b\left(m^c\right) = c \cdot \log_b(m)$ for any $b > 0$, $b \neq 1$, and $m, n > 0$

The first property of logarithms states $\log_b(m \cdot n) = \log_b(m) + \log_b(n)$. For example, $\log_2(4) + \log_2(8)$ asks to find the exponent to which 2 is raised to in order to get a result of 4, and the exponent 2 is raised to in order to get a result of 8, and then add those two exponents together $(2 + 3 = 5)$. We could have instead considered $\log_2(4 \cdot 8) = \log_2(32)$ and determined directly that 5 is the exponent 2 must be raised to in order to get a result of 32.

---

*8. Use your understanding of logarithmic functions (knowledge of what the input and output quantities represent) to justify at least one of the logarithmic properties above.

*9. Use the properties of logarithms to simplify the following expressions.

a. $\ln(x) + \ln(x)$

b. $\log_3(5) + \log_3(2)$

c. $\log_4(16) - \log_4(4)$

d. $\log_3(2) + \log_5(3)$

e. $\log_3(9) + \log_3(4) - \log_3(6)$

f. $\log_7(49) - \log_7(-3)$

*10. Solve each of the following for x.

a. $\log_5(30x^2) = 3$

b. $\log(1.5x) = 0$

c. $\log_3 x + \log_3(2x) = 3$

11. Let $f(x) = 10^x$ and $g(x) = \log(x)$.

a. Complete the tables.

x	$f(x)$
-2	
-1	
0	
1	
2	

x	$g(x)$
0.01	
0.1	
1	
10	
100	

b. Evaluate the following expressions:

i. $g(f(-2))$

ii. $g(f(0))$

iii. $f(g(1))$

iv. $f(g(10))$

v. $f(g(x))$

vi. $g(f(x))$

c. What do you notice about the relationship between the functions f and g?

12. Use the fact that the word "log" is the name of a function and the statement $\log_2(x)$ represents the output values of the logarithmic function (with a base of 2) for varying values of x.

a. What does $\log_2(x)$ represent?

b. What does $\log_4(x)$ represent?

c. Write an equation relating $\log_2(x)$ and $\log_4(x)$.

d. Write an equation relating $\log_3(y)$ and $\log_{27}(y)$.

*1. Solve each of the following equations for x. Find the exact answer and then use your calculator to approximate the answer to the nearest thousandth (3 decimal places).

  a. $4 = 3^x$

  b. $22.4 = 17.5(3.4)^x$

*2. a. Describe using your own words what the output of $\log_2(A)$ represents? What does the output of $\log_2(B)$ represent? If $A = B$, how are $\log_2(A)$ and $\log_2(B)$ related?

  b. Why is it mathematically valid to "take the log" of both sides of an equation? (*In other words, how do you know the two sides of the equation are still equal?*)

  c. Is it mathematically valid to take the log of both sides of an equation if the expressions on each side of the equation can have a value less than or equal to zero? Explain.

3. In 2009 the enrollment at Mainland High School was 1650 students. Administrators predict that the enrollment will increase by 2.2% each year.

  a. Determine when the enrollment at Mainland High School is expected to reach 2600 students.

  b. Suppose that Mainland High School requires 5 teachers per 65 students. How many teachers are required if the student enrollment is 2600 students?

c. Define a function *f* that gives the number of years since 2009 in terms of the number of teachers employed by Mainland High School.

*4. *Recall the alpaca problem from Module 4 Investigation 4.* A farm in Canada had 218 alpacas on January 1, 2004. After 8 years the alpaca population decreased to 187. Assume the number of alpacas decays exponentially.
a. Define a function *f* that expresses the number of alpacas on the farm in terms of the number of years since 2004.

b. Using the function defined in part (a) estimate the number of alpaca on the farm in 2013.

c. Use algebraic methods to determine the year when the number of alpaca on the farm was 165. Check your answer using a graph or by using the function you defined in part (a).

*5. *Recall the total number of COVID-19 cases in Arizona context from Module 4 Investigation 4.* The number of total COVID-19 cases in Arizona increased from 16,891 on May 23, 2020 to 31,264 on June 11, 2020 [a period of 20 days]. Assume the total number of cases increased exponentially over this period.
a. Define a function *f* that expresses the total number of Arizona COVID-19 cases during this period in terms of the number of days since May 23, 2020 (or recover the function you defined in that investigation).

b. How many days after May 23, 2020 did the total number of COVID-19 cases in Arizona reach each of the following totals? [*You may assume the trend continued past 20 days if necessary.*]
    i. 20,000 total cases   ii. 40,000 total cases

c. Define a function *h* that determines the number of days since May 23, 2020 in terms of the total number of Arizona COVID-19 cases.

d. How is function *f* defined in part (a) above related to function *h* defined in part (c) above?

*6. In 1990, the population of Diamond Bar was 45,000 and the population of Chino Hills was 37,000. In 1992, the population of Diamond Bar was 51,750 and the population of Chino Hills was 44,400. Assuming both cities' populations grow at an exponential rate, how many years will it take for the population of Chino Hills to surpass the population of Diamond Bar?

*7. In 1935, Charles Richter, a seismologist at the California Institute of Technology, invented a method for comparing the magnitude of earthquakes. Since the amplitude of seismic waves of different earthquakes can be over a million times as large as the waves of small earthquakes, Richter developed a special scale (now called the Richter Scale) for measuring the magnitude of these earthquake waves. This formula or function takes as input the amplitude of the seismic waves of an earthquake and outputs what we call the magnitude of the earthquake. The magnitude is the Richter scale rating of the strength of the earthquake and is the number reported in the media.

The magnitude M of an earthquake whose seismic waves are of amplitude A is defined to be

$$M = \log_{10}\left(\frac{A}{A_0}\right),$$

where A_0 represents the amplitude of seismic waves of a "standard" earthquake (whose amplitude is 1 micron = 10^{-4} cm).

a. i. Describe the meaning of the ratio $\dfrac{A}{A_0}$ in this context.

   ii. Describe the effect of inputting $\dfrac{A}{A_0}$ into the $\log_{10}$ function and what the value of M represents.

b. Let M_1 and M_2 represent the magnitudes of two earthquakes whose seismic waves are of amplitudes A_1 and A_2, respectively. Use the formula given above and the properties of logarithms to define a simplified formula for the difference $M_2 - M_1$ in terms of A_1 and A_2.

---

c. On May 22, 1960, a 9.5 magnitude earthquake struck near Valdivia, Chile. On November 1, 2015, a 4.1 magnitude earthquake struck Phoenix, Arizona. How many times as large were the seismic waves of the Chile earthquake compared to the Arizona earthquake?

d. Write a formula that determines the amplitude of the seismic waves of an earthquake in terms of the magnitude of the earthquake.

8. On May 15, 2016, Nancy paid the balance on her loans and was finally debt free! Since she had extra money, she deposited $50 into a savings account that same day and then forgot about the account (so she made no additional deposits). The number of dollars $w(t)$ in this savings account t months since May 15, 2016 is given by the function w defined by, $w(t) = \log_4(t+1) + \log_2(t+1) + 50$.

   a. What is the practical domain and range of the function w?

   b. Evaluate $w(15)$. What does this value represent in the situation?

   c. Evaluate $w(43) - w(0)$. What does this value represent in the situation?

   d. What does $\frac{w(22) - w(0)}{22 - 0}$ represent in this situation?

   e. Solve for t when $w(t) = 114$.

   f. Define a function, p, that expresses the number of months since May 15, 2016 in terms of the amount of money a in Nancy's savings account.

I. PERCENTAGES AND PERCENT CHANGE (TEXT: S1)

1. You worked a total of 450 minutes at your job yesterday.
   a. What amount of time corresponds to 78% of the total time you spent working yesterday?
   b. If you had worked for 78% of the 450 minutes, what percent of the total time do you still have left to work? What is this time in minutes?

2. Suppose that you fill a glass with 860 milliliters of water.
   a. What volume of water corresponds to 44% of 860 mL?
   b. If you drank 44% of the water you poured into the glass, what percent of the starting 860 mL remains? How much water is this in mL?

3. A local bicycle shop placed older model bikes on sale last weekend. A customer bought a bike on sale for $299 that had a normal retail price of $495.

   sale price: $299

   retail price: $495

   a. How many times as large is the sale price compared with the retail price?
   b. *Fill in the blank*: "The sale price is _____% of the retail price."
   c. How many times as large is the retail price compared with the sale price?
   d. *Fill in the blank*: "The retail price is _____% of the sale price."

4. Your doctor recommends that you increase or decrease your daily intake of certain vitamins and minerals. For each recommendation described, do the following.
   > i) State the percent change in daily intake.
   > ii) State the number by which we can multiply the original daily intake to find the new daily intake.
   > iii) Find the new recommended daily intake amount.
   a. original vitamin D intake: 3µg; increase: 60%
   b. original potassium intake: 1500mg; increase: 135%
   c. original zinc intake: 30mg; decrease: 45%
   d. original calcium intake: 820mg; increase: 25%

5. A store is adjusting the prices of several items. For each price change, do the following.
   > i) State the percent change.
   > ii) State the number by which we can multiply the old price to find the new price.
   > iii) Find the new price.
   a. old price: $29; increase: 5%
   b. old price: $84; increase: 140%
   c. old price: $89.99; decrease: 34%
   d. old price: $6.49; decrease: 22%

6. A major record label has seen its annual operating profit decrease in recent years, likely because of greater accessibility of music online. In 2011, the label's operating profit was $135 million. By 2015, the label's operating profit had decreased by 43% (a percent change of –43%).
   a. What was the record label company's operating profit in 2015?
   b. The record label wants to increase its operating profit to $100 million by 2017. By what percent must the label's operating profit increase from its 2015 value to reach $100 million within the next two years?

7. During a recession, Tina had to take a 10% pay cut. Her original salary was $58,240.
   a. What was her salary after the pay cut?
   b. Once the recession was over, Tina's company wanted to increase her pay to her original salary. What percent change in her salary was required to return it to its original $58,240?

II. COMPARING LINEAR AND EXPONENTIAL BEHAVIOR (TEXT: S2)

8. Joni, a graduating biomedical engineer, was offered two positions, one with Company A, and the other with Company B. Company A offered her a starting salary of $66,000 with a 5% guaranteed raise at the end of each of the first five years. Company B offered her $75,000 as her starting salary with a guaranteed raise of $1,500 every year for the first 5 years. She likes both companies and believes she will continue with the company she selects for at least 5 years.
   a. What is the ratio of Joni's salary one year compared to her salary in the previous year for Company A? Describe how to interpret this ratio.
   b. Fill in the blank: If Joni works for Company A, her salary for any year is _____% of her salary for the previous year.
   c. Define a function f that expresses her salary at Company A in terms of the number of years n since she accepts the position.
   d. Define a function g that expresses her salary at Company B in terms of the number of years n since she accepts the position.
   e. After how many years will Joni's salary at Company A overtake her salary at Company B?
   f. Suppose that she will work for one of these companies for exactly 5 years. A classmate says she should choose Company A because by the time she leaves the company she will have a higher salary. Do you agree? Defend your reasoning.
   g. Construct a graph of the two functions and explain the meaning of the intersection point in the context of this situation (Note that the graphs of the two functions will not be continuous. Before creating your graph think about the discrete instances when her salary changes and the fact that her salary remains constant during each year.)
   h. Find and interpret the following using the functions created in parts (b) and (c).
      i. $f(8)$ ii. $g^{-1}(8200)$ iii. $g(16)$ iv. $f(20) - f(11)$

9. You are researching jobs in advertising.
   a. You are told that the salary for Job A increases exponentially. The salary for Year 1 is $29,000 while the salary for Year 2 will be $31,066.
      i. How many times as large is the salary in Year 2 compared to the salary in Year 1?
      ii. Fill in the blank: The salary in Year 2 is _____% of the salary in Year 1.
      iii. What is the percent change in salary from Year 1 to Year 2?
      iv. If the salary continues to increase by the same percent each year, what will the salary be in Year 5?
   b. You are told that the salary for Job B also increases exponentially. The salary for Year 1 is $27,500 while the salary for Year 2 will be $28,600.
      i. How many times as large is the salary in Year 2 compared to the salary in Year 1?
      ii. Fill in the blank: The salary in Year 2 is _____% of the salary in Year 1.
      iii. What is the percent change in salary from Year 1 to Year 2?
      iv. If the salary continues to increase by the same percent each year, what will the salary be in Year 5?
   c. Based on the information found in parts (a) and (b), which job would you take? Explain your reasoning.

10. A chemist monitored the mass of bacteria in a Petri dish after applying a chemical to kill the bacteria. This chemical causes the mass of bacteria of this type to decrease by 12% each hour that elapses after applying the chemical. The mass when applying the chemical was 203 micrograms.
   a. Fill in the blank. At any given time, the mass of the bacteria is _____% of the mass one hour earlier.
   b. What would the mass of bacteria be after 1 hour? After 2 hours? After 8 hours?

Exercise continues on the next page.

c. Define a function that expresses the mass of bacteria B remaining in the Petri dish as a function of the number of hours t since applying the chemical.

d. After how many hours since applying the chemical will the bacteria's mass be less than one microgram?

e. A different type of chemical was applied to another Petri dish of bacteria that caused the mass to decrease by 26% each hour. If the initial mass of bacteria in this dish was 230 micrograms, define a function which gives the mass of bacteria in this Petri dish as a function of the number of hours t that have elapsed since applying the chemical.

11. Last year, Jenny invested her birthday money in 3 different penny stocks. The following functions represent the daily value (rounded to the nearest cent) of each stock over the first 7 days after making the investment. For each investment,
    i) State the amount of the initial investment;
    ii) Describe how the value of the investment grew over the first 7 days since making the investment?
    iii) Determine the value of investment at the end of the 7th day. (Round your answers to the nearest penny.)
    a. $h(n) = 25(1.45)^n$
    b. $f(n) = 120(3)^n$
    c. $g(n) = 275(0.9)^n$

12. The rabbit population on a 10-acre wildlife preserve was 24 on January 1, 2011.
    a. Assuming the number of rabbits doubled each year, determine a function that gives the number of rabbits R in the preserve in terms of the number of years t elapsed since January 1, 2011.
    b. Using the function created in part (a), approximate the number of rabbits in the preserve on January 1, 2018.
    c. What is the percent change per year if the population of rabbits doubles each year?

13. A firework stand opened on June 28th. The number of customers at the stand over the first 5 days since it was opened is defined by the function, $N(x) = 7(2)^x$. At the end of July 2, the function was updated to $N(x) = 224(3)^x$, with x continuing to represent the number of days passed since the firework stand was opened. What implications can you draw from this information?

14. When a teacher asked her beginning algebra class to provide an example of an exponential function, over half of the students offered the function $f(x) = x^2$ because "the function is growing faster and faster".
    a. What is your assessment of these students' answer?
    b. Compare the growth patterns of $f(x) = x^2$ and $g(x) = 2^x$. Construct a table of values and create a graph of f and g on the same axes. Then use the graphs and table values to compare and contrast their growth patterns.
    c. Compare the growth patterns of $h(x) = x^3$ and $j(x) = 3^x$. Construct a table of values and create a graph of h and j on the same axes. Then use the graphs and table values to compare and contrast their growth patterns.

15. Define a function that models each town's population growth in terms of the number of years since the town was established.
    a. Smallsville starts with 500 people and grows by 10 people per year.
    b. Growsville starts with 500 people and grows by 10% each year.
    c. Shrinktown starts with 500 people and declines by 10% each year.
    d. Littletown starts with 500 people and declines by 10 people per year.

16. Each given function defines the population for a city in terms of the time t in years since the city was established. Write a sentence that describes the city's initial population and growth pattern.

    a. $f(t) = 2000(1.24)^t$ b. $g(t) = 1500 + 20t$ c. $h(t) = 4000(0.68)^t$

    d. $k(t) = 2500 - 40t$ e. $f(t) = 1500(1.4)^{t/2}$

17. The US population was about 273.6 million in 1996. Since that time the population increased by approximately 1.1% each year.

    a. Define a function f that expresses the population P of the US in millions as a function of the number of years t since 1996.

    b. What was the approximate population of the US in 2010 according to your model?

    c. Assuming the population continues to grow at 1.1% per year, in what year will the US population reach 400 million people?

18. A biologist counted 426 bees in a bee colony. His tracking of the colony revealed that the number of bees increased by 4% each month over the next year.

    a. Approximately how many bees were in the colony 3 months after the initial count?

    b. How many months passed before the number of bees reached 500?

19. A company purchases a new car for $25,000 for their employees to use. For accounting purposes, they decide to depreciate the value of the car by 14.5% each year.

    a. Using this method of depreciation, what is the value of the car after 2 years?

    b. What is the ratio of the car's value in one year compared to its value the previous year? Explain the meaning of the value of this ratio.

    c. Define a function that models the value of the car as a function of the number of years since the company purchased it.

    d. When will the value of the car be less than $1000?

III. 1-Unit Growth and Decay Factors, Percent Change, and Initial Values (Text: S2)

20. Determine the growth or decay factor, percent change, and initial value for each of the following exponential functions.

    a. $f(x) = 2^x$

        Initial Value:
        1-unit Growth Factor:
        1-unit Percent Change:

    b. $f(x) = (0.98)^x$

        Initial Value:
        1-unit Decay Factor:
        1-unit Percent Change:

    c. $f(x) = 0.56 \cdot (0.25)^x$

        Initial Value:
        1-unit Decay Factor:
        1-unit Percent Change:

    d. $f(x) = 3 \cdot (1.6)^x$

        Initial Value:
        1-unit Growth Factor:
        1-unit Percent Change:

21. The given tables represent patterns of exponential growth. Determine the initial value, 1-unit growth/decay factor, 1-unit percent change, and define a formula to model the data in each table.

    a.

x	0	1	2	3
$f(x)$	512	384	288	216

    Initial Value: 1-unit Decay Factor:
    1-unit Percent Change: Formula:

    b.

x	1	2	3
$g(x)$	11.2	15.68	21.952

    Initial Value: 1-unit Growth Factor:
    1-unit Percent Change: Formula:

22. Determine the growth or decay factor, percent change, initial value, and formula for each of the following graphs modeled by exponential functions.

a.

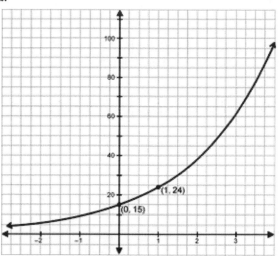

b.

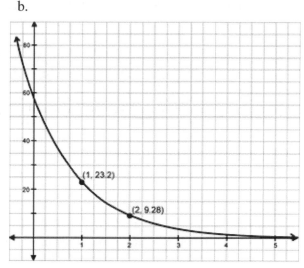

Growth Factor:

Percent Change:

Formula:

Decay Factor:

Percent Change:

Formula:

23. An investment of $9000 decreased by 2.4% each month.
    a. Determine the initial value of the investment, the 1-month decay factor, and the 1-month percent change of the investment.
    b. Define a function *f* that determines the value of the investment in terms of the number of months since the initial investment was made.
    c. Determine the value of the investment 12 months after the money was invested.

24. The population of Canada in 1990 was 27,512,000 and in 2000 it was 30,689,000. Assume that Canada's population increased exponentially over this time.
    a. Determine the initial value, the 10-year growth factor, and the 10-year percent change of Canada's population.
    b. Define a function *f* that defines the population of Canada in terms of the number of *decades* (10-year periods) since 1990.
    c. Assuming Canada's growth rate remains consistent, predict the approximate population of Canada in 2030.

25. One year after an investment was made, the amount of money in the account was $1844.50. Two years after the investment was made, the amount of money in the account was $2001.28.
    a. Determine the initial value of the investment, the investment's 1-month growth factor, and the investment's 1-month percent change.
    b. Define a function *f* that defines the value of the investment in terms of the number of months since the investment was made.
    c. Determine when the value of the investment will reach $4000.

26. The mass of bacteria in a Petri dish was initially measured to be 14 micrograms and increased by 12% each hour over a period of 15 hours.
    a. Determine the 1-hour growth factor of the bacteria.
    b. By what total percent did the bacteria increase during the 15-hour time period?
    c. Determine how long it will take for the mass of bacteria to double.

IV. PARTIAL AND *N*-UNIT GROWTH FACTORS (TEXT: S3)

27. A certain strain of bacteria that is growing on your kitchen counter doubles in number every 5 minutes. Assume that you start with only one bacterium.
    a. What quantities are changing in this situation? What quantities are not changing?
    b. Define a function that gives the number of bacteria present after *n* 5-minute intervals.
    c. Represent the number of bacteria present after 30 minutes. How many bacteria are there at that time?
    d. How many bacteria are present after 1 minute? 2 minutes? 5 minutes?
    e. Determine the 1-minute growth factor. By what percent does the bacteria change every *minute*?
    f. Determine the 2-minute growth factor and 2-minute percent change.
    g. Define a function that gives the number of bacteria present after *t* minutes.
    h. Define a function that gives the number of bacteria present after *h* hours.

28. The number of deer in a park reserve was counted to be 27 on January 1, 2000. Six months later, the number of deer was counted to be 38.
    a. Determine the 6-month growth factor and 6-month percent change for the situation.
    b. Determine the 1-month growth factor and 1-month percent change for the situation.
    c. Define a function that relates the number deer in the park reserve since January 1, 2000 in terms of the number of months that have elapsed since January 1, 2000. (*Assuming the number of deer continues to increase by the same percent each month.*)
    d. Define a function that relates the number deer in the park reserve since January , 2000 in terms of the number of *6-month intervals* that have elapsed since January 1, 2000. (*Assuming the number of deer continues to increase by the same percent each month.*)
    e. Use one of your functions to determine when the number of deer will reach approximately 125.

29. The population of Egypt in 2002 was 73,312,600, and was 78,887,000 in 2006. Assume the population of Egypt grew exponentially over this period.
    a. Determine the 4-year growth factor and percent change.
    b. Determine the 1-year growth factor and percent change.
    c. Define a function that gives the population of Egypt in terms of the number of years that have elapsed since 2002.
    d. Re-write your function from part (c) so that it gives the population of Egypt in terms of the number of *decades* that have elapsed since 2002.
    e. Assuming Egypt's population continued to grow according to this model, how long will it take for population of Egypt to double?

30. An animal reserve in Arizona had 93 wild coyotes. Due to drought, there were only 61 coyotes after 3 months. Assume that the number of coyotes decreases (or decays) exponentially.
    a. Find the 3-month decay factor and percent change.
    b. Find the 1-month decay factor and percent change.
    c. Define a function that represents the number of coyotes in terms of the number of elapsed months.
    d. Re-write your function from part (c) so that it gives the number of coyotes in terms of the number of *years* that have elapsed.
    e. How long will it take for the number of coyotes to be one-half of the original number?

31. Sales of music albums decreased by 6% each year over a period of 8 years.
    a. Determine the decay factor for 1 year.
    b. By what total percent did album sales change during the 8-year time period?
    c. How long will it take for the album sales to be half of what they were at the beginning of this period? A*ssume the trend continues in the future.*

---

32. The population of a town increased or decreased by the following percentages. For each situation, find the population's *annual percent change.*
    a. increases by 60% every 12 years
    b. decreases by 35% every 7 years
    c. doubles in size every 6 years
    d. increases by 4.2% every 2 months
    e. decreases by 7% every week

33. The number of asthma sufferers in the world was about 84 million in 1990 and 130 million in 2001. Let N represent the number of asthma sufferers (in millions) worldwide t years after 1990.
    a. Define a function that expresses N as a linear function of t. Describe the meaning of the slope and vertical intercept in the context of the problem.
    b. Define a function that expresses N as an exponential function of t. Describe the meaning of the growth factor and the vertical intercept in the context of the problem.
    c. The world's population grew by an annual percent change of 3.7% over those 11 years. Did the percent of the world's population who suffer from asthma increase or decrease?
    d. What is the long-term implication of choosing a linear vs. exponential model to make future predictions about the number of asthma sufferers in the world?

34. In the second half of the 20th century, the city of Phoenix, Arizona exploded in size. Between 1960 and 2000, the population of Phoenix increased by 2.76% each year. In the year 2000, the population of Phoenix was determined to be 1.32 million people.
    a. Define an exponential function f that gives the population P in Phoenix (in millions of people) where the input values t represent the number of years after the year 2000.
    b. Sketch a graph of this function.
    c. What input to your function will give the population of Phoenix in 1972? 1983? 1994? According to your model, approximately how many people lived in Phoenix in 1972? 1983? 1994?
    d. In what year did the population reach 1,000,000 people?
    e. The change in the population of Phoenix is increasing for equal changes in time. Illustrate this on the graph of f for at least 3 different equal intervals of time.
    f. Now, define an exponential function, g, modeling the population of Phoenix in millions of people n years after 1960. Sketch a graph of this function.
    h. How does the function you created in part (f) compare to the original function created in part (a)? How are the graphs of the two functions similar and different? Explain your reasoning.

V. *n*-UNIT GROWTH AND DECAY FACTORS ($n \neq 1$) (TEXT: S3)

35. For each function given below, find the initial value, specified factors, and specified percent changes.
    a. $f(x) = 2^{x/3}$
       i) Initial Value:
       ii) 1-unit Growth Factor:
       iii) 1-unit Percent Change:
       iv) 4-unit Growth Factor:
       v) 1/5-unit Growth Factor:

    b. $f(x) = 5 \cdot (0.98)^{1/4}$
       i) Initial Value:
       ii) 1-unit Decay Factor:
       iii) 1-unit Percent Change:
       iv) 3-unit Decay Factor:
       v) 1/2-unit Decay Factor:

    c. $f(x) = (0.25)^{2x}$
       i) Initial Value:
       ii) 1-unit Decay Factor:
       iii) 1-unit Percent Change:
       iv) 5-unit Decay Factor:
       v) 1/3-unit Decay Factor:

    d. $f(x) = 3 \cdot (1.6)^{5x}$
       i) Initial Value:
       ii) 1-unit Growth Factor:
       iii) 1-unit Percent Change:
       iv) 3-unit Growth Factor:
       v) 1/4-unit Growth Factor:

36. For each exponential relationship, determine the specified growth factors. Write your answers in exponential and decimal form (rounded to the nearest thousandth). In addition, determine the initial value and define the exponential function formula that models the data.

a.

x	1	3	5	7
$f(x)$	512	384	288	216

   i) Initial Value:
   ii) 1-unit Decay Factor:
   iii) 1-unit Percent Change:
   iv) 3-unit Decay Factor:
   v) 1/3-unit Decay Factor:
   vi) Function Formula:

b.

x	2	5	8	11
$g(x)$	8	11.2	15.68	21.952

   i) Initial Value:
   ii) 1-unit Growth Factor:
   iii) 1-unit Percent Change:
   iv) 2-unit Growth Factor:
   v) 1/4-unit Growth Factor:
   vi) Function Formula:

37. Determine the specified decay factors. Write your answers in exponential and decimal form (round to the nearest thousandth). Then determine the function formula that models the data.
   a. 3-unit Decay Factor:
   b. 5-unit Decay Factor:
   c. 1-unit Decay Factor:
   d. 0.6-unit Decay Factor:
   e. Initial Value:
   f. Function Formula:

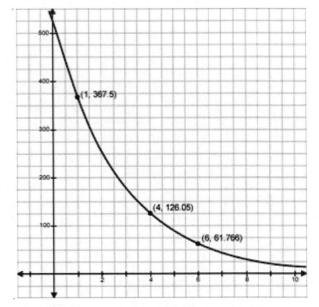

38. Determine the specified decay factors. Write your answers in exponential and decimal form (round to the nearest thousandth). Then determine the function formula that models the data.
   a. 2-unit Growth Factor:
   b. ½-unit Growth Factor:
   c. 1-unit Growth Factor:
   d. 2.5-unit Growth Factor:
   e. 5-unit Growth Factor:
   f. Initial Value:
   g. Function Formula:

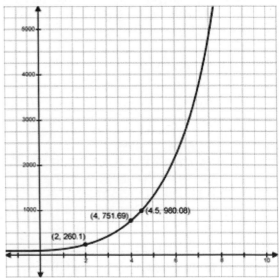

39. *Determine the specified growth factors for the following relationship. Write your answers in exponential and decimal form (round to the nearest thousandth). The mass of bacteria in an experiment at time t days after its start is given by* $f(t) = 95(4)^{t/5}$.
    a. 5-day growth factor: b. 1-day growth factor:
    c. Select the statement(s) below that describes the behavior of the function above and give reasoning for why the statement(s) are true:
       i. An initial mass of 95 μg of bacteria quadruples every 1/5-day.
       ii. An initial mass of 95 μg of bacteria quadruples every 5-days.
       iii. An initial mass of 95 μg of bacteria increases by 50% every 1-day.

40. *Determine the specified growth factors for the following relationship. Write your answers in exponential and decimal form (round to the nearest thousandth). The number of buffalo in a wildlife preserve at time t months after its initial measure is given by* $f(t) = 49(0.97)^{2t}$.
    a. 1-month decay factor: b. ½-month decay factor:
    c. Select the statement(s) below that describe(s) the behavior of the function above:
       i. An initial number of 49 buffalo decreases by 3% every ½-day.
       ii. An initial number of 49 buffalo decreases by 3% every 2-days.
       iii. An initial number of 49 buffalo decreases by 5.91% every 1-month.

41. *Determine the specified growth factors for each relationship. Write your answers in exponential and decimal form (round to the nearest thousandth).*
    a. The population of Jackson has 57,421 people and is growing by 4.78% every 4 years.
       i) 8-year growth factor:
       ii) What is the percent change (increase) over 8-year period?
       iii) 1-year growth factor:
       iv) What is the percent change (increase) over 1-year period?
    b. The amount of drug B in your body decreases by 18% every 3 hours.
       i) 24-hour decay factor:
       ii) What is the percent change (decrease) over a 24-hour period?
       iii) 1-hour decay factor:
       iv) What is the percent change (decrease) over a 1-hour period?

42. For each relationship, do the following.
    a. Determine whether the relationship could be linear or exponential.
    b. Fill in the blanks of the tables based on your answer to (a).
    c. Define a function formula that models the data in each table.

Table i		Table ii		Table iii	
Input	Output	Input	Output	Input	Output
−2.0	−10.0	−4.0	0.1111	0.0	132.0
−1.5		−3.0		0.5	
−1.0	−7.5	−2.0	0.3333	1.0	29.04
−0.5		−1.0		1.5	
0.0	−5.0	0.0	1.0	2.0	6.3888
0.5		1.0		2.5	
1.0	−2.5	2.0	3.0	3.0	1.4055
1.5		3.0		3.5	
2.0	0	4.0	9.0	4.0	0.309
				4.5	

VI. COMPOUNDING PERIODS & COMPOUND INTEREST FORMULA (TEXT: S4)

43. The given table illustrates the value of an investment from the end of the 7th compounding period to the end of the 10th compounding period.

Number of compounding periods p	Investment value
7	$7,105.57
8	$7,209.31
9	$7,314.57
10	$7,421.36

    a. Verify that the data in the table represents exponential growth.

    b. How does the investment value change from the end of the 7th to the end of the 9th compounding period? From the end of the 8th to the end of the 10th compounding period?

    c. What is the value of the investment after 3 compounding periods? After 20 compounding periods?

    d. Define a function f that expresses the investment as a function of the number of compounding periods p. Note: p is defined to be the values $\{0,1,2,3,...\}$. Explain the meaning of each of the values in your function.

    e. Construct a graph of the function defined in part (d). Label the axes appropriately, then pick one coordinate point from the graph and explain what it represents.

    f. Determine the number of compounding periods until the investment reaches $400,000.

44. For each of the following accounts, determine the percent change per compounding period. Give your answer in both decimal and percentage form.

    a. 5% APR compounded monthly   b. 6.7% APR compounded quarterly

    c. 3.2% APR compounded daily   d. 8% APR compounded each hour

45. For each of the accounts in Exercise #44, give the growth factor per compounding period, then give the annual growth factor and the annual percent change (APY).

46. If you invest $1,200 in a CD with an APR of 3.5% compounded monthly, the following expression will calculate the value of the CD in 6 years: $1200\left(1+\frac{0.035}{12}\right)^{12(6)}$

    Explain what each part of the expression represents in this calculation.

    a. $\frac{0.035}{12}$ b. $1+\frac{0.035}{12}$ c. $12(6)$ d. $\left(1+\frac{0.035}{12}\right)^{12}$ e. $\left(1+\frac{0.035}{12}\right)^{12(6)}$

47. Write an expression that would calculate the value of the following account after 14 years: $8100 is invested at an APR of 4.8% compounded semiannually (twice per year).

48. Write an expression that would calculate the value of the following account after 30 years: $16,000 is invested at an APR of 3.5% compounded daily.

49. An investment of $10,000 with Barnes Bank earns a 2.42% APR compounded *monthly*.

    a. Define a function that gives the investment's value as a function of the number of years since it began.

    b. Determine the investment's value after 20 years.

    c. Determine the annual growth factor and annual percent change (APY).

    d. Determine how long will it take the investment value to double.

    e. Another bank says they will pay you the same interest (2.42% APR), but compounded daily. What would be the value of a $10,000 investment with this bank, leaving it there for 20 years? Compare this answer to your account with Barnes Bank.

50. You just won $1000 in the lottery and you decide to invest this money for 10 years.
    a. Which of the three different accounts would you choose to invest your $1000? Provide calculations for each account and justify your reasoning.
        - Account #1 pays 14% interest each year, compounded annually (once per year).
        - Account #2 pays 13.5% interest per year, compounded monthly.
        - Account #3 pays 13% interest per year, compounded weekly.
    b. Describe how your money increases as time increases for each account. Be sure to incorporate the annual growth factor and annual percent change into your description!

51. Karen received an inheritance from her grandparents and wants to invest the money. She is offered the following options of accounts to invest into:
    - 4.5% APR, compounded semi-annually
    - 4.3% APR, compounded daily
    a. Which option should she choose? Explain your reasoning using the accounts' APY.
    b. If Karen decides on the first option, by what *percent* will her investment have *increased* after 10 years?

52. John decides to start saving money for a new car. He knows he can invest money into an account which will earn a 6.5% APR, compounded weekly, and would like to have saved $10,000 after 5 years.
    a. How much money will he need to invest into the account now so that he has $10,000 after 5 years?
    b. Determine the APY (Annual Percent Yield) for the account.
    c. Determine the 5-year percent change for the account.

VII.   INVESTMENT ACTIVITY: FOCUS ON FORMULAS AND MOTIVATING *e* (TEXT: S4)

53. $4500 is initially invested into an account with an APR of 7%.
    a. Determine the value of the account at the end of the given years and the APY for each type of compounding per year.
    b. Compare the short-term vs. the long-term impact of increasing the number of times interest is compounded each year.

Year	Compounded Monthly	Compounded Daily	Compounded Continuously
1			
2			
4			
10			
30			
APY			

54. Determine the value of each of the following accounts after 14 years.
    a. Initial investment is $4000 with a 7.1% APR compounded continuously.
    b. Initial investment is $750 with a 3.3% APR compounded continuously.
    c. Initial investment is $2000 with a 5% APY compounded continuously.

55. $2000 is initially invested in an account with an APR of 3.4% compounded continuously.
    a. Determine the value of the account at the end of 5 years.
    b. Write a function that models the value of the account at the end of *t* years.
    c. What is the annual percent change (APY) of the account? What does this value represent?
    d. What is the percent change over 10 years for this account?

56. $18,000 is initially invested in an account with an APR of 5.1%.
    a. Does it make a bigger impact going from compounding interest annually to monthly or going from compounding interest daily to continuously?
    b. What is the annual percent change (APY) for each of the four compounding methods listed in (a)?
    c. If interest is compounded continuously, how much interest does the account earn over the first 10 years?

57. An initial population of 32 million people increases at a continuous percent rate of 1.9% per year since the year 2000.
    a. Determine the function that gives the population in terms of the number of years since 2000.
    b. Determine the population in the year 2019.
    c. What is the annual growth factor for this context? Explain two ways we can determine this value.
    d. What does your answer to part (c) represent in this context?

58. Carbon-14 is used to estimate the age of organic compounds. Over time, carbon-14 decays at a continuous percent rate of 11.4% per thousand years from the moment the organism containing it dies. Carbon-14 is typically measured in micrograms.
    a. What quantities are changing in this situation? What quantities are not changing?
    b. Define a function that gives the amount, at any moment, of Carbon-14 remaining in a piece of wood that starts out with 150 micrograms of Carbon-14. (*Remember that Carbon-14's decay rate is per thousand years.*)
    c. Construct a graph of this function. Be sure to label your axes!
    d. What is the percent change every 1000 years?
    e. What is the percent change every 5000 years? Explain your reasoning.

VIII. THE INVERSE OF AN EXPONENTIAL FUNCTION (TEXT: S5)

Log Property #1: $\log_b(m \cdot n) = \log_b(m) + \log_b(n)$ for any $b > 0$, $b \neq 1$ and $m, n > 0$

Log Property #2: $\log_b\left(\frac{m}{n}\right) = \log_b(m) - \log_b(n)$ for any $b > 0$, $b \neq 1$ and $m, n > 0$

Log Property #3: $\log_b(m^c) = c \cdot \log_b m$ for any $b > 0$, $b \neq 1$ and $m, n > 0$

59. Estimate the value of each logarithmic expression. Explain why your estimate makes sense:
    a. $\log_4(60)$
    b. $\log_3(143)$
    c. $\log_8\left(\frac{1}{64}\right)$
    d. $\log_9(27)$

60. Find the unknown in each of the following equations. *Estimate if necessary. For parts (g) and (h), assume z is a positive number.*
    a. $\log_3(20) = y$
    b. $2(\log_5(625)) = y$
    c. $\log_5(1) = y$
    d. $\log_2(30) = y$
    e. $\log_2(x) = 4.2$
    f. $\log_7(x) = -2$
    g. $\log_z(343) = 3$
    h. $\log_z(1) = 0$

61. For each of the following:
    • Write the expression as a single logarithm.
    • Evaluate to a single number or estimate the value of the expression.
    a. $\log_5(6.25) + \log_5(100) + \log_5(25)$
    b. $\log_4\left(\frac{1}{32}\right) + \log_4\left(\frac{1}{8}\right)$
    c. $2 \cdot \log_6(12) + \log_6(4)$
    d. $\ln(6) + \ln(3) - \ln(2)$
    e. $\log_5(6) - \log_5(100)$
    f. $\frac{1}{2}(\log_5(8) + \log_5(3))$

62. Solve each of the following for x. (*Hint: Use the properties discussed in class or your understanding of logarithms.*)

    a. $\log(0.01x) = 0$ b. $\log_5(25x^2) = 6$ c. $\ln\left(\frac{x}{10}\right) = 4$

63. Solve each of the following for x.

    a. $\log_2(2+x) + \log_2(7) = 3$ b. $\ln(3x^2) - \ln(5x) = \ln(x+9)$

64. Rewrite each of the following as sums and differences of a logarithm of some number.

    a. $\log_7\left(\frac{4}{y}\right)$ b. $\log_2\left(x^4 \cdot 12\right)$ c. $\log_5\left(\frac{10x^3}{y^5}\right)$

65. Rewrite each of the following exponential equations in logarithmic form.

    a. $y = 11^x$ b. $y = 1.7(3.2)^t$ c. $y = 200(1.0027)^{12t}$

66. Graph each of the following functions.

    a. $f(x) = \log_3(x)$ b. $g(x) = \log_5(x)$

67. a. Sketch the following two functions on the same set of axes: $f(x) = \log(x)$ and $g(x) = \log_5(x)$

    b. Consider the rates of change of each of the functions. Explain why $g(x)$ has a greater rate of change than $f(x)$ for all values of $x > 1$.

    c. For what values of x is $\log(x) > \log_5(x)$? Explain your reasoning.

IX. SOLVING EXPONENTIAL AND LOGARITHMIC EQUATIONS (TEXT: S5)

68. The amount of an investment is represented by $f(t) = 4186.58(1.025)^t$. Algebraically, determine when (in compounding periods, t) the investment will reach $1,000,000.

69. An initial amount of 120 mg of caffeine is metabolized in the body and decreases at a continuous percent rate of 21% per hour.

    a. Define an exponential function that gives the amount of caffeine remaining in the body after t hours.

    b. How many hours will it take for the amount of caffeine to reach half of the initial amount? (*Solve this both graphically and symbolically to verify your answers.*)

70. An initial investment of $6000 is made to an account with an APR of 4.7%.

    a. If interest is compounded monthly, how many years will it take for the account balance to be $10,290.32? Solve algebraically and check your answer.

    b. If interest is compounded continuously, how many years will it take for the account balance to be $9,510.15?

71. The amount of medicine in a patient's bloodstream for reducing high blood pressure decreases at a continuous percent rate of 27% per hour. This medicine is effective until the amount in the bloodstream drops below 1.2 mg. A doctor prescribes a dose of 85 mg.

    a. Define the function A that models the amount of medicine remaining in the bloodstream after t hours.

    b. About how long until 45 mg of medicine remains in the patient's bloodstream? 1.2 mg?

72. The rate at which a wound heals can be modeled by the exponential function $f(n) = Ie^{-0.1316n}$ where I represents the initial size of the wound in square millimeters and $f(n)$ represents the size of the wound after n days. This function assumes no infection is present and no antibiotic ointment is used to speed healing.
    a. Suppose you scrape your knee and get a wound 300 square millimeters in size. Define a function to model the size of the wound with respect to time.
    b. How large will the wound be after one week?
    c. How long will it take for the wound to be 20% of its original size?
    d. You want to know how long it will take to reduce the size of the wound to 20% of the size you determined in part (c). How will the amount of time it takes to do this compare to the amount of time it took to reduce the wound to 20% of its original size? Explain your reasoning.

73. Given the function $f(t) = 10(0.71)^t$, where t is in years, complete the following.
    a. What is the annual percent change?
    b. Convert the function $f(t) = 10(0.71)^t$ into the equivalent form $f(t) = ae^{kt}$
    c. What is the continuous annual percent rate?

74. The town of Gilbertville increased from a population of 3,562 people in 1970 to a population of 9,765 in 2000.
    a. Define an exponential function that models the town's population as a function of the number of years since 1970.
    b. What is the annual percent change?
    c. Use your function to predict the town's population in 2019.
    d. According to your function, when will the town's population reach 40,000 people? (*Answer this question both graphically and symbolically.*)
    e. After how many years from *any* reference year will the population triple?

This investigation contains review and practice with important skills and procedures you may need in this module and future modules. Your instructor may assign this investigation as an introduction to the module or may ask you to complete select exercises "just in time" to help you when needed. Alternatively, you can complete these exercises on your own to help review important skills.

Factoring Variable Expressions
Use this section prior to the module or with/after Investigation 1.

Factoring is the process of rewriting a number as a product of factors. For example, we can rewrite 10 as $5 \cdot 2$. The value is still the same but we've written the number as a product.

What if the number is the value of a varying quantity? For example, the variable x could be used to represent all the values of some varying quantity. Then, for ALL possible values of x, the expression $(\frac{1}{2}x)(2)$ has the same value but is written as a product.

If $x = 8$, then $\frac{1}{2}x = 4$, and so $(\frac{1}{2}x)(2)$ is $(4)(2)$, which is 8 written as a product of two numbers.

1. Repeat this for each of the following values of x (use the expression $(\frac{1}{2}x)(2)$ to write that value of x as a product of two numbers).
    a. $x = 12$ b. $x = 17$ c. $x = -28$ d. $x = 1.3$

2. a. Can you create a variable expression different from $(\frac{1}{2}x)(2)$ that will always represent the value of x written as a product of two other numbers? Give an example.

    b. Choose two different values for x and show how your answer to part (a) "works".

3. $2x^2 + 3x$ is a variable expression that can be written in factored form as $x(2x + 3)$.
    a. Pick two values for x and evaluate $2x^2 + 3x$.

    b. Use the same values for x to show that $x(2x + 3)$ represents the same value as $2x^2 + 3x$ but written as a product of two numbers.

In Exercises #4-6, do the following.
    (i) Rewrite the expression in factored form.
    (ii) Pick a value for x and use it to evaluate the original expression and your factored form to demonstrate that your answer to part (a) represents the same value as the original expression written as a product of two numbers.

4. $5x + 35$ 5. $6x^2 - 13x$ 6. $3x^2 + 12x$

In Exercises #7-9 you are given a variable expression written in factored form. Use the distributive property to rewrite each expression in expanded form.

7. $6x(x + 7)$ 8. $6y(3y - 2x)$ 9. $a(ab^2 + a^3c)$

Expanding Binomial Products

Use this section prior to the module or with/after Investigation 3.

When a product involves two variable expressions with two or more terms, writing the expanded form can sometimes be tricky. One very useful way to think about the product is to use an area model. To demonstrate the idea, let's think about the number 21, which could be written as a product $(7)(3)$, and even broken down further and written as a product of two sums like $(3 + 4)(2 + 1)$. Let's visualize this product as representing a rectangular area of 21 square units. *See diagram to the right.*

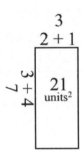

If we break down this area into sub-rectangles, we can see that the entire area is made up of four sub-areas measuring $(3)(2)$ square units, $(3)(1)$ square units, $(4)(2)$ square units, and $(4)(1)$ square units. *See diagram to the left.*

Thus, $(3 + 4)(2 + 1)$ is equivalent to $(3)(2) + (3)(1) + (4)(2) + (4)(1)$.

We can use the same basic idea to visualize the product of $(x + 2)(x + 3)$. *See diagram to the right.*

Using this diagram it should be clear that $(x + 2)(x + 3)$ is equivalent to $x^2 + 3x + 2x + (2)(3)$. This expression can be simplified to $x^2 + 5x + 6$. This is the expanded (and simplified) form.

In Exercises #10-12, rewrite each product in expanded form by first drawing an area diagram.

10. $(x + 6)(x + 2)$

11. $(x + 3)(x + 5)$

12. $(x + 1)(x + 4)$

In Exercises #13-18, rewrite each product in expanded form. *You do not need to draw an area diagram but you can if it helps you. Remember that something like $(x + 3)^2$ is shorthand for $(x + 3)(x + 3)$.*

13. $(x + 8)(x + 3)$

14. $(x + 1)(2x + 5)$

15. $(2x + 3)(3x + 7)$

16. $(x + 5)^2$

17. $(x + 8)^2$

18. $(2 + x)^2$

In Exercises #19-24, rewrite each product in expanded form. *We recommend rewriting subtraction as addition with a negative to avoid sign errors. For example, rewrite $x - 5$ as $x + (-5)$. You do not need to draw an area diagram but you can if it helps you. Remember that something like $(x - 3)^2$ is shorthand for $(x - 3)(x - 3)$.*

19. $(x - 3)(x + 5)$

20. $(x + 2)(3x - 1)$

21. $(3x - 2)(3x - 4)$

22. $(x - 1)^2$

23. $(x - 6)^2$

24. $(2x - 3)^2$

25. a. Your friend said that $(a + b)^2$ can be written as $a^2 + b^2$. Is he right? Justify your answer using an area diagram.

   b. Use an area diagram to help you write the general expanded form for any product $(a - b)^2$.

Factoring Polynomial Expressions
Use this section prior to the module or with/after Investigation 3.

If we want to write a polynomial expression like $x^2 + 7x + 10$ in factored form we can use an area diagram to help us organize our thinking. From our work above it should be clear that the top left rectangle in our area diagram will have an area of x^2 square units (with side lengths of x and x units) and the bottom right rectangle must have an area of 10 square units.

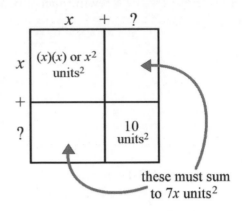

What we need to figure out are the dimensions of the bottom right rectangle that will 1) make its area 10 square units and 2) the areas of the other two rectangles sum to $7x$ square units.

Our options include any pair of numbers with a product of 10 (which includes $10 \cdot 1$ and $5 \cdot 2$). These options are shown below.

$$(x+10)(x+1) \qquad \text{or} \qquad (x+5)(x+2)$$

26. a. Which of the options is "correct"? What does that tell us about the factored form of $x^2 + 7x + 10$?

b. Choose a value for x and evaluate the expression $x^2 + 7x + 10$. Then use the same value to evaluate the factored form to show that it represents the same value of $x^2 + 7x + 10$ but written as a product of two numbers.

c. Repeat part (b) with a different value for x.

In Exercises #27-32, rewrite each expression in factored form. For example, $x^2 + 9x + 14$ would be rewritten as $(x+7)(x+2)$. *Draw an area diagram if it helps you complete the process.*

27. $x^2 + 8x + 12$

28. $x^2 + 9x + 8$

29. $x^2 + 2x - 15$

30. $x^2 - x - 42$

31. $x^2 - 7x + 12$

32. $x^2 + 3x - 40$

In Exercises #33-35, rewrite each expression in factored form by first factoring x out of the expression. For example, $x^3 - 3x^2 - 10x$ can be written as $x(x^2 - 3x - 10)$, and then can be written as $x(x+2)(x-5)$.

33. $x^3 + 10x^2 + 24x$

34. $x^3 - 6x^2 - 16x$

35. $x^3 - 6x^2 + 9x$

Solving Polynomial Equations
Use this section prior to the module or with/after Investigation 4.

Consider the polynomial equation $x^2 + 8x + 7 = 0$. The solutions are the values of x such that the expression $x^2 + 8x + 7$ has a value of 0. There are several ways to solve these equations. Let's start by factoring the expression.

$$x^2 + 8x + 7 = 0$$
$$(x+7)(x+1) = 0$$

We now have a product of two numbers (the number $x + 7$ and the number $x + 1$) that evaluates to 0. What does that tell us? Well, the only way a product of two numbers can be 0 is if one of the two numbers in the product is 0. In other words, if $ab = 0$, then either $a = 0$ or $b = 0$. This is known as the ***zero-product property***.

What that means in this context is that either $(x + 7) = 0$ or $(x + 1) = 0$.

$$x^2 + 8x + 7 = 0$$
$$(x+7)(x+1) = 0$$
$$x + 7 = 0 \quad \text{or} \quad x + 1 = 0$$
$$x = -7 \quad \text{or} \quad x = -1$$

So either $x = -7$ or $x = -1$. These are the only values that could make $x^2 + 8x + 7$ have a value of 0.

36. Substitute $x = -7$ and $x = -1$ into the expression $x^2 + 8x + 7$ to show that each produces a value of 0.

In Exercises #37-39 solve each equation by factoring and using the zero-product property. Check your answer by graphing. *To do this, graph a function f whose outputs are defined by the polynomial expression and demonstrate that the function value is 0 for each x-value solution.*

37. $x^2 - 13x + 30 = 0$ 38. $x^2 + 7x - 18 - 0$ 39. $2x^2 + 5x + 3 = 0$

Sometimes the equation is not set equal to 0, meaning we can't use the zero-product property. However, we can rewrite the equation so that the zero-product property applies. See the following example.

$$x^2 + 9x + 13 = 4x + 7 \qquad \text{(subtract } 4x \text{ and 7 from both sides)}$$
$$x^2 + 5x + 6 = 0$$
$$(x+3)(x+2) = 0$$
$$x + 3 = 0 \quad \text{or} \quad x + 2 = 0$$
$$x = -3 \quad \text{or} \quad x = -2$$

In Exercises #40-42, solve each equation by first rewriting it so that the equation is equal to 0.

40. $x^2 + 5x + 2 = 3x + 10$ 41. $x^2 - 12x + 24 = -2x + 3$ 42. $x^2 + 1 = 2x + 16$

If the equation is quadratic (and written in the form $ax^2 + bx + c = 0$), then the x-values that make the expression evaluate to 0 (the solutions to the equation) can be represented by $x = -\frac{b}{2a} \pm \frac{\sqrt{b^2 - 4ac}}{2a}$. This is called the **quadratic formula** and represents the solutions to equations of the form $ax^2 + bx + c = 0$.

Note that "$\pm$" indicates that the quadratic formula can produce two solutions.

$$2x^2 - 5x + 1 = 0 \quad \text{(so } a = 2, \, b = -5, \, c = 1\text{)}$$

$$x = -\frac{b}{2a} \pm \frac{\sqrt{b^2 - 4ac}}{2a}$$

$$x = -\frac{-5}{2(2)} \pm \frac{\sqrt{(-5)^2 - 4(2)(1)}}{2(2)}$$

$$x = -\frac{-5}{4} \pm \frac{\sqrt{25 - 8}}{4}$$

$$x = \frac{5}{4} \pm \frac{\sqrt{17}}{4}$$

The solutions are $x = \frac{5}{4} + \frac{\sqrt{17}}{4}$ and $x = \frac{5}{4} - \frac{\sqrt{17}}{4}$.

In Exercises #43-48, solve each equation using the quadratic formula $x = -\frac{b}{2a} \pm \frac{\sqrt{b^2 - 4ac}}{2a}$. *Be very careful and pay attention to the signs.*

43. $2x^2 + 6x + 3 = 0$ 44. $x^2 + 4x - 1 = 0$ 45. $x^2 - 3x - 5 = 0$

46. $3x^2 + 8x + 3 = 0$ 47. $5x^2 - 2x - 1 = 0$ 48. $2x^2 + 11x - 7 = 0$

*1. Examine the bottle given below and imagine the bottle filling with water. While imagining the bottle filling with water consider how the height of the water in the bottle changes.

a. Use this reasoning to draw a rough sketch of a graph to represent the height of the water in terms of the volume of water in the bottle. *Label your axes. You may add units of your choosing (e.g., cups and cm.) or ignore the numbers on the axes. Your graph does not need to be exact—a rough sketch is good enough!*

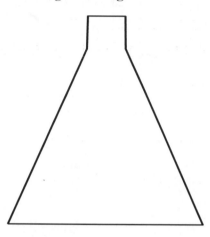 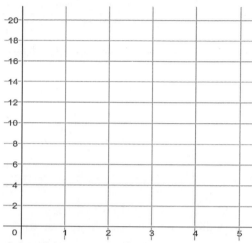

b. Based on your graph, describe how the height of the water in the bottle changes as the volume of water in the bottle increases throughout the domain.

*2. a. Now, use the applet provided in the PowerPoint to determine the height of the water in the bottle for the 7 different volume values noted on the applet (and appearing in the table).

Note that it is useful to select values of the volume by repeatedly adding the same volume of water for each section of the bottle when determining the water's height. Explain why this strategy is helpful.

Volume of Water (cups)	Height of Water (cm)	Change in the Water's Height (cm)
0	0	
1	2	
2		
3		
4		
4.2		
4.4		

b. Use the values in your table to sketch a more accurate graph of the height of the water in the bottle in terms of the volume of water in the bottle.

*3. The given graph represents the *height of water in another bottle* in terms of the *volume of water in the bottle*. Use this graph to answer the following questions.

a. Consider how the water's height changes as the volume of water increases from 0 to 2.5 cups, and 2.5 to 5 cups.
   i. On which of these two intervals does the water's height increase more? Explain.

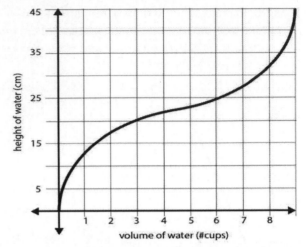

   ii. What does this information suggest about the shape of the bottom half of the bottle?

b. Consider how the water's height changes as the volume increases from 5 to 7.5 cups, and from 7.5 to 10 cups.
   i. On which interval does the water's height increase more? Explain.

   ii. What does this information suggest about the shape of the upper part of the bottle?

c. Note that the volume-height graph changes from curving downward to curving upward when 5 cups have been added to the bottle. What might this graphical information convey about the bottle's shape?

d. Draw a picture of a bottle that would produce the volume-height relationship conveyed in the above graph. Label landmarks on your bottle, and on the graph in part (a), where the function's behavior changes in important ways.

The given graph (from Investigation 1) represents the relationship between the height (in cm) and volume of water (in cups) in a bottle. Call this function relationship f.

*1. a. On the graph, plot the points $(0, f(0))$ and $(5, f(5))$.

    b. Draw a line that passes through these points.

    c. Determine the slope of the line you drew by estimating values for $f(0)$ and $f(5)$ on the graph.

    d. What does this slope represent in the context of this problem?

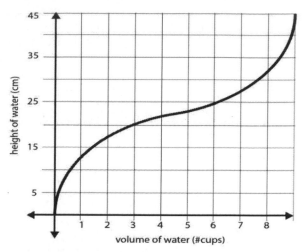

It's clear that, as the volume of water in the bottle changes from 0 cups to 5 cups, the water height does NOT change at a constant rate with respect to its volume. However, the slope of the line connecting the two points represents *a constant rate of change,* of the water height with respect to the water's volume, that would achieve the same net change in the water height, that was achieved by the actual function, for the same change in volume. This is the ***function's average rate of change over that interval.***

    e. It is often useful to determine the ***constant rate of change*** (*average rate of change*) of a function over smaller and smaller intervals of the function's domain. Discuss with your classmates why you think this might be useful. Provide your answer below.

    f. Construct 5 lines on the graph of f by connecting the points:
      i) $(0, f(0))$ and $(1, f(1))$;   ii) $(1, f(1))$ and $(2, f(2))$;   iii) $(2, f(2))$ and $(3, f(3))$;
      iv) $(3, f(3))$ and $(4, f(4))$ v) $(4, f(4))$ and $(5, f(5))$

    g. Discuss the following questions with your classmates, then provide you answer.

      i. how does the *function's average rate of change* (over 1 cup intervals of the function's domain) ***change***, as the volume of water in the bottle increases from 0 to 5 cups.

      ii. What does the pattern of change in these values tell you about the shape of the function's graph?

      iii. What does the pattern of change in these values tell you about how the height of the water in the bottle and volume of water in the bottle are changing together?

*2. A 10-foot ladder is leaning against a wall at a vertical position. A carpenter then pulls the base of the ladder away from the wall at a constant speed until the top of the ladder hits the floor.

a. After considering the situation and discussing with your classmates, make a conjecture about the speed at which the top of the ladder falls toward the floor. Does it fall at a constant speed, or does it slow down, or speed up? Justify your claim.

b. Construct a function f, that defines the distance of the top of the ladder from the floor, $f(x)$, in terms of the distance of the bottom of the ladder from the wall, x. (Hint: Use Pythagorean Theorem.)

c. Use f to complete the following:

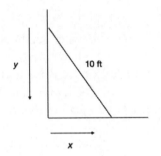

Distance of the bottom of the ladder from the wall, x.	Distance of the top of the ladder from the floor $y = f(x)$	Change in the distance of the top of the ladder from the floor, $\Delta f(x)$
0	10	
2		
4		
6		
8		
10	0	

d. For each x-interval specified in the given table, determine f's average rate of change.

e. Examine the patterns of change in these AROC values and explain what this conveys about how the top of the ladder falls toward the floor. Does it fall at a constant speed, or does it slow down, or speed up? Explain.

x- interval	AROC of $f(x)$ with respect to x.
0-2	
2-4	
4-6	
6-8	
8-10	

3. For each of the given functions, find the average rate of change over the indicated interval and then discuss with a partner or as a class how to interpret the meaning of your answer.

a. $f(x) = x^3 - 6x^2 + 9x + 10$

as x changes from $x = -1$ to $x = 3$

b. $g(x) = -3 \cdot \sqrt{x+4} + 6$

as x changes from $x = -4$ to $x = 5$

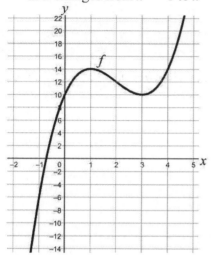
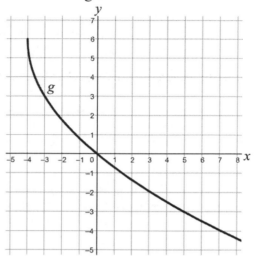

*4. a. Illustrate a change in x from –6.3 to –3.9 on the number line below.

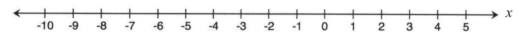

b. T or F: The value of x is decreasing as x changes from –6.3 to –3.9. Justify your response.

c. Function h is defined by $h(x) = -x^2$.

i. As x increases from –4 to –2.5, $h(x)$ changes from _____ to _____.

ii. As x increases from –2.5 to –1, $h(x)$ changes from _____ to _____.

iii. Determine the average rate of change of h on the interval from $x = $ 4 to 2.5 and explain how to interpret the meaning of your answer.

iv. Determine the average rate of change of h on the interval from $x = -2.5$ to –1 and explain how to interpret the meaning of your answer.

v. **T or F:** The average rate of change of h on the interval from –4 to –2.5 is greater than the average rate of change of h on the interval from –2.5 to –1.

Rates of change (and average rates of change) are measurements of how a function's output quantity changes in tandem with changes in the input quantity. **Concavity** is a measurement of how a function's rate of change itself changes in tandem with changes in a function's input quantity.

Concavity

For any function f, imagine looking at an interval of the domain from $x = a$ to $x = b$ and dividing it into any number of equal-sized subintervals.

- f is said to have **positive concavity** on the interval (a, b) if the function's average rate over successive intervals always increases.
- f is said to have **negative concavity** on the interval (a, b) if the function's average rate over successive intervals always decreases.

---

*5. Return to Exercise #1. Over what interval(s) of the domain does the function have positive concavity? Over what interval(s) of the domain does the function have negative concavity? Make sure you can justify your answer.

In Exercise #1 the function changed from having negative concavity when the volume was less than 5 cups to having positive concavity when the volume was greater than 5 cups. The point $(5, 23)$ is thus an **inflection point** for the function.

Inflection Point

A function f has an inflection point at $(a, f(a))$ if the function has a different concavity when $x < a$ compared to when $x > a$.

---

6. For each of the given functions, do the following.
    i. State any interval(s) of the domain over which it appears the function has positive concavity.
    ii. State any interval(s) of the domain over which it appears the function has negative concavity.
    iii. Estimate any inflection points for the function.

*a. $f(x) = x^3 - 6x^2 + 9x + 10$

b. $g(x) = -3 \cdot \sqrt{x+4} + 6$

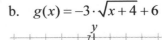

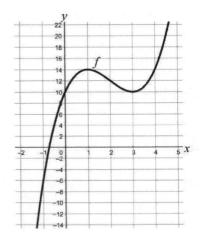

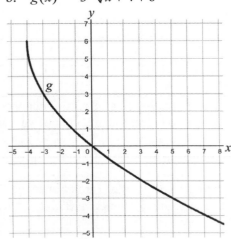

You have likely heard the phrase ***concave up*** and ***concave down*** instead of ***positive concavity*** and ***negative concavity***, especially when discussing graphs. Graphs of functions over intervals with positive concavity "curve up" as x increases. Graphs of functions over intervals with negative concavity "curve down" as x increases. Examples are shown below, along with line segments whose slopes represent the functions' average rates of change over consecutive intervals.

Example I: positive concavity ("concave up") on the entire domain

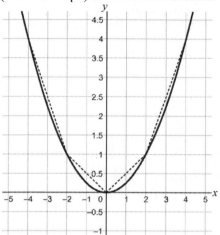

Example II: positive concavity ("concave up") on the entire domain

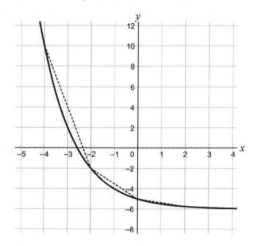

Example III: negative concavity ("concave down") on the entire domain

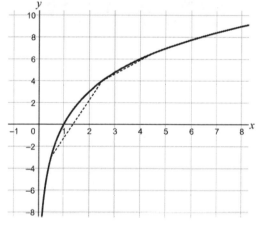

Example IV: negative concavity ("concave down") on the entire domain

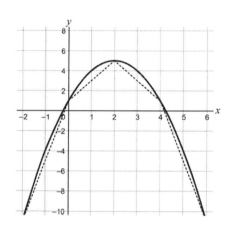

7. *a. For Example I explain why the function has positive concavity over its entire domain (from what we can tell). Then explain how you can "see" the graph curving upwards as x increases.

    b. For Example II explain why the function has positive concavity over its entire domain (from what we can tell). Then explain how you can "see" the graph curving upwards as x increases.

    *c. For Example III explain why the function has negative concavity over its entire domain (from what we can tell). Then explain how you can "see" the graph curving down as x increases.

    d. For Example IV explain why the function has negative concavity over its entire domain (from what we can tell). Then explain how you can "see" the graph curving down as x increases.

8. Return to the graphs in Exercise #6. Identify the interval(s) over which the function graphs are "concave up" and "concave down".

9. Graph each of the following functions using a graphing calculator or graphing software. For each, do the following.
    i. Identify the interval(s) over which the function's graph is "concave up". *It's okay to estimate.*
    ii. Identify the interval(s) over which the function's graph is "concave down".

    a. $f(x) = x^3$

    b. $g(x) = -2x^2 + 4x - 1$

    c. $h(x) = -x^3 + 6x$

    d. $j(x) = 3^x$

10. At 5:00 pm Karen started walking from the grocery store back to her house.

    a. Fill in the table by determining the number of feet Karen is from home, *d*. Then use the information in the table to determine Karen's average rate of change (of her distance from home in terms of the number of minutes since she started walking) on each specified interval

Change in the number of minutes since Karen started walking Δt	Number of minutes since Karen started walking t	Number of feet Karen is from home d	Change in the number of feet Karen is from home Δd	Average rate of change of Karen's distance with respect to time
	0	118		
			−1.5	
	0.5			
			−3.2	
	1			
			−6.5	
	1.5			
			−7.1	
	2			

    b. Sketch a graph of the number of feet Karen is from home in terms of the number of minutes since Karen started walking. (*Be sure to label your axes.*)

    c. Does this function have positive concavity ("concave up"), negative concavity ("concave down"), or some combination of both on the interval $0 < t < 2$? Make sure you can justify your answer.

    d. Describe how the quantities *number of minutes since Karen started walking* and the *number of feet Karen is from home* change together.

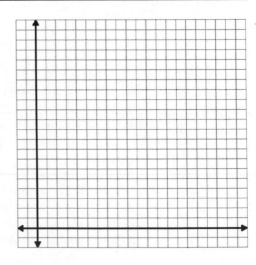

*1. Regal Theatre puts on community theatre productions each season. After several years of gathering data, they noted some trends between the length of the plays they produced and the number of tickets they sold for Friday evening shows. They created the model shown below.

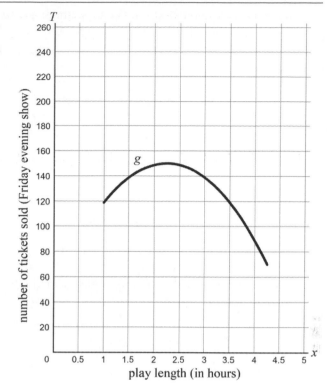

play length (hours) *x*	number of tickets sold to Friday evening shows (according to the model) *g(x)*
1	119
1.5	139
1.75	145
2	149
2.25	150
2.5	149
3	139
3.5	119
4	89

a. What does the ordered pair (3, 139) represent in the context of this situation?

b. Determine the function's average rate of change over each of the following intervals. Make sure to describe how to interpret the meaning of the average rate of change in each case.
   i. $1 \le x \le 2.25$ ii. $2.5 \le x \le 4$

c. Does the function have positive concavity or negative concavity (or a combination of the two)? What does this tell you about the situation?

*2. Regal Theatre contacted a similar theatre in another city (Player Productions). Player Productions said their model was virtually identical to the model created by Regal Theatre. The only difference is that they sold 30 fewer tickets each Friday night compared to Regal Theatre.
   a. Complete the table of values for Player Productions's model.

   b. Describe how the outputs of functions g and h compare for any input value x.

   c. Draw a graph of h on the axes given in Exercise #1.

   d. How does the average rate of change of h on the intervals $1 \leq x \leq 2.25$ and $2.5 \leq x \leq 4$ compare to the average rate of change of g on the same intervals? Why?

play length (hours) x	number of tickets sold to Friday evening shows (according to the model) $h(x)$
1	
1.5	
1.75	
2	
2.25	
2.5	
3	
3.5	
4	

   e. Use function notation to express the outputs of h in terms of the outputs of g.

*3. A third theatre company (Actor's Guildhouse) found a similar trend, however they sell 1.5 times as many tickets to their Friday shows compared to the Regal Theatre.
   a. Complete the given table of values for Actor's Guildhouse's model.

   b. Describe how the outputs of functions g and f compare for any input value x.

   c. Draw a graph of f on the axes given in Exercise #1.

   d. How does the average rate of change of f on the intervals $1 \leq x \leq 2.25$ and $2.5 \leq x \leq 4$ compare to the average rate of change of g on the same intervals? Why?

play length (hours) x	number of tickets sold to Friday evening shows (according to the model) $f(x)$
1	
1.5	
1.75	
2	
2.25	
2.5	
3	
3.5	
4	

   e. Use function notation to express the outputs of f in terms of the outputs of g.

*4. The number of tickets to a Friday performance in terms of the play length (in hours) for a fourth theatre company (Stage Left) is modeled by function j.

a. If $j(x) = g(x - 0.5)$, how do the ticket sales for a Friday performance compare at Stage Left and Regal Theatre? [*Hint: Which function needs a larger input value to produce the same output value?*] How would the graphs of the functions compare?

b. If instead $j(x) = 2g(x) - 15$, how do the ticket sales for a Friday performance compare at Stage Left and Regal Theatre? How would the graphs of the functions compare?

*5. The graph of h is given. Draw the graph of g if $g(x) = -h(x + 3)$.

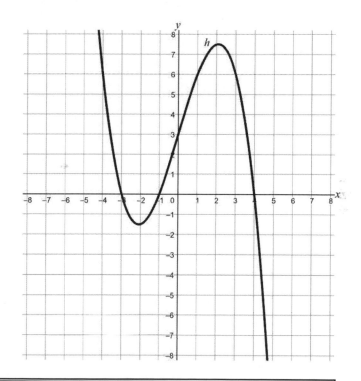

Many of the functions we have worked with in this module (and that we will continue to work with in the next few investigations) are *polynomial functions*.

Polynomial Functions

When expressed as a formula, a polynomial function can be written in the form
$f(x) = a_n x^n + a_{n-1} x^{n-1} + a_{n-2} x^{n-2} + \ldots + a_2 x^2 + a_1 x + a_0$ where n, $n - 1$, etc. are natural numbers and a_n, a_{n-1}, etc. are real numbers. *Note that the coefficients could have a value of zero.*

Examples of polynomial function formulas include $f(x) = 4x^3 + 8.2x^2 - 3x + 1.7$ and $g(x) = -5x^6 + 13x^2$. Quadratic functions and linear functions are examples of polynomials.

---

*6. Define each function in terms of the other.

   a.   i. Express the outputs of g in terms of the outputs of f .

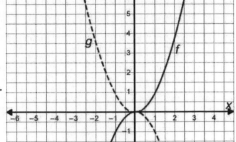

      ii. Express the outputs of f in terms of the outputs of g .

   b.   i. Express the outputs of g in terms of the outputs of f .

      ii. Express the outputs of f in terms of the outputs of g .

   c.   i. Express the outputs of g in terms of the outputs of f .

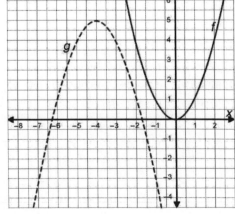

      ii. Express the outputs of f in terms of the outputs of g .

   d. Which of the functions in parts (a) through (c) could be polynomial functions?

*1. Use your graphing calculator to sketch a graph of the following quadratic functions and compare the behavior and properties of these functions. For each of these functions determine the roots (x-intercepts) of the function.

a. $f(x) = x^2$

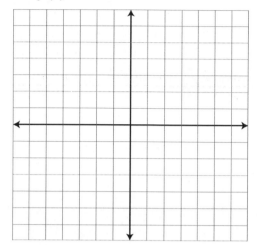

b. $g(x) = 3x^2$

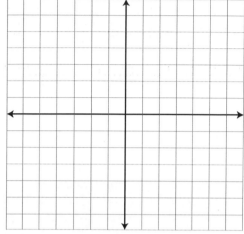

c. $h(x) = 3x^2 - 27$

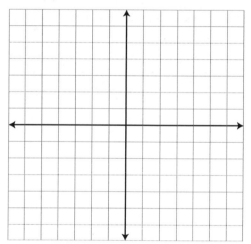

d. $k(x) = -3(x-7)^2$

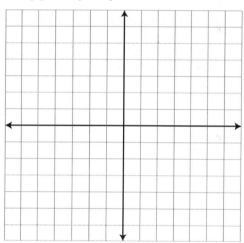

e. $p(x) = (2x-9)(x+4)$

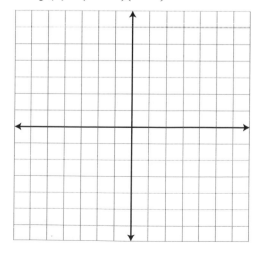

*2. Use your graphs from Exercise #1 to answer the following questions.
   a. When a quadratic function *f* is given in factored form explain why the roots (*x*-intercepts) occur
       where each factor has a value of 0. [*Note, if a function is given in factored form such as*
       $f(x) = (x-3)(x+4)$, *the factors are* $x-3$ *and* $x+4$.]

   b. Is it possible for a quadratic function to have positive concavity over one interval of its domain
       and negative concavity over some other interval of its domain? Explain.

   c. Does a quadratic function always have a maximum value?

   d. How do you determine if a quadratic function will have positive or negative concavity?

   e. What is the relationship between the *x*-coordinate where the maximum/minimum function value
       occurs and the function's roots?

*3. The function *f* is defined by $f(x) = (x+7)(x-3)$.
   a. Use algebraic methods to determine the roots of *f*.

   b. Determine the *x*-coordinate of *f* 's minimum value (also referenced as the *x* coordinate of the
       vertex of the parabola generated by graphing *f*). Then find the function's minimum value.

   c. Find the average rate of change of $f(x)$ with respect to *x* as *x* increases from 2 to 7.4.

   d. Evaluate $f(3)$.

It's relatively easy to determine the vertex and real roots (if any) of a quadratic function given in factored form. The roots are the values of x that make one of the factors 0, and the x-value of the vertex is directly in between the roots.

When the function is instead in **standard form** (in the form $f(x) = ax^2 + bx + c$ where a, b, and c are real numbers) the **quadratic formula** can be used to determine the function's roots.

The Quadratic Formula

For a quadratic function of the form $f(x) = ax^2 + bx + c$ where a, b, and c are real numbers*, the **quadratic formula** represents the function's roots (the values of x such that $f(x) = 0$).

$$x = \frac{-b}{2a} \pm \frac{\sqrt{b^2 - 4ac}}{2a}$$

Note that b or c could be 0.

---

4. Use the quadratic formula to determine the roots of each of the following functions (find the exact values and decimal approximations). Then use a graphing calculator to check your work. *It might help to first list the values of a, b, and c to use in the formula.*

   *a. $f(x) = 2x^2 + 5x + 1$

   b. $f(x) = x^2 - 7x + 5$

*5. a. What does $x = \frac{-b}{2a}$ represent in the context of a quadratic function?

   b. What does $f\left(\frac{-b}{2a}\right)$ represent in the context of a quadratic function f?

   c. What does $\pm \frac{\sqrt{b^2 - 4ac}}{2a}$ represent in the context of a quadratic function f?

   d. If $\frac{\sqrt{b^2 - 4ac}}{2a} = 0$ for some quadratic function f, what does this tell us about the function?

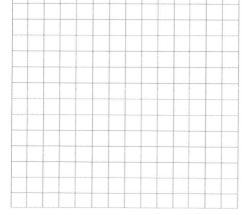

   e. Assume that for a given quadratic function, f, the solutions to the quadratic formula are $x = -4 \pm 3.25$. Create a possible sketch of this quadratic function and illustrate the vertex and roots of this quadratic function.

---

*6. An ice cream shop finds that its weekly profit P (measured in dollars) as a function of the price x (measured in dollars) it charges per ice cream cone is given by the function k, defined by
$k(x) = -125x^2 + 670x - 125$ where $P = k(x)$.

a. Determine the maximum weekly profit and the price of an ice cream cone that produces that maximum profit.

b. If the cost of the ice cream cone is too low then the ice cream shop will not make a profit. Determine what the ice cream shop needs to charge in order to break even (make a profit of $0.00).

c. If the cost of the ice cream cone is too high then not enough people will want to buy ice cream. As a result the weekly profit will be $0.00. Determine what the ice cream shop would have to charge for this to happen (the profit to be $0.00).

d. The profit function for Cold & Creamy (another ice cream shop) is defined by the function g where $g(x) = k(x - 2)$. Does the function g have the same maximum value as k? What is the price per ice cream cone that Cold & Creamy ice cream shop must charge to produce a maximum profit? Explain.

e. Describe the meaning of $h(x) = k(x) - 125$ and then compare/contrast the maximum values for each function.

7. Define the function formula that generates a parabola that has horizontal intercepts at $x = 3$ and $x = -5$, and passes through the point (2, 3).

*1. The function f is defined by $f(x) = (x+1)(x-3)^2$.

    a.   Use algebraic methods to find the roots (x-intercepts) of f.

    b.   What do the roots of a polynomial function represent?

    c.   Explain why a polynomial function's roots occur where one of the factors has a value of zero.

    d.   Draw a number line representing values of x and highlight each of the following:
        i.   the intervals of x where f's output is positive

        ii.   the intervals of x where f's output is negative

    e. Construct a rough sketch of the graph of f.

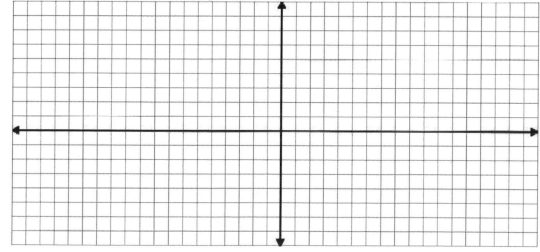

2.   How do the functions f and h compare, given that $f(x) = (x+1)(x-3)^2$ and $h(x) = 2(x+1)(x-3)^2$?

*3. How do the functions h and k compare, given that $h(x) = 2(x+1)(x-3)^2$ and $k(x) = -2(x+1)(x-3)^2$?

*4. Let $f(x) = x^3$, $g(x) = x^5$, $h(x) = x^2$, $j(x) = x^8$. (A function of the form $p(x) = ax^n$, given that a and n are real numbers, is called a **power function**. Note that n must be a non-negative integer for p to be a polynomial function.)

    a.   Describe how each power function varies as:

        i. x increases without bound (also written $x \to \infty$)

           $x \to \infty, \quad f(x) \to$ _____ $x \to \infty, \quad g(x) \to$ _____

           $x \to \infty, \quad h(x) \to$ _____ $x \to \infty, \quad j(x) \to$ _____

       ii. x decreases without bound (also written $x \to -\infty$)

           $x \to -\infty, \quad f(x) \to$ _____ $x \to -\infty, \quad g(x) \to$ _____

           $x \to -\infty, \quad h(x) \to$ _____ $x \to -\infty, \quad j(x) \to$ _____

    b.   What general statements can you make about how the exponent on a power function impacts the behavior of the function?

    c.   What changes if each function changes to have a coefficient of –2 (for example, $f(x) = x^3$ changes to become $f(x) = -2x^3$)?

*5. Given $f(x) = x^3 + 4x^2 - 6x - 12$ and $g(x) = x^3$, do the following.

    a.   Graph the functions with a window size ranging from $x = -5$ to $x = 5$ and from $y = -20$ to $y = 20$. Do the functions have similar behavior?

    b.   Graph the functions with a window size ranging from $x = -50$ to $x = 50$ and from $y = -10,000$ to $y = 10,000$. Do the functions have similar behavior?

    c.   If your answers are different to parts (a) and (b), why are they different?

> ### A Polynomial Function's Leading Term
>
> For a polynomial function f represented by a formula, the leading term is the term containing the largest exponent. For example, given $f(x) = 5x^4 + 7x^3 + 5x + 100$, the leading term is $5x^4$, and given $g(x) = 5 - 4x^3$, the leading term is $-4x^3$.

It's important to be able to identify the leading term for a polynomial function because the leading term dictates the function's **end behavior** (the behavior of $f(x)$ as $x \to \pm\infty$, or as x increases or decreases without bound). This is because of the power of repeated multiplication with very large numbers – larger exponents create huge discrepancies in the relative size of each term in the long-run.

For example, consider the function $f(x) = 5x^4 + 7x^3 + 5x + 100$. $f(10,000) = 50,007,000,000,050,100$, but how does each term contribute to this function value?

- $5(10,000)^4 = 50,000,000,000,000,000,000$, which is 99.986% of the value of $f(10,000)$
- $7(10,000)^3 = 7,000,000,000,000$, which is 0.014% of the value of $f(10,000)$
- $5(x) = 50,000$, which is 0.0000000001% of the value of $f(10,000)$
- 100 is 0.0000000000002% of the value of $f(10,000)$

As x continues to increase without bound, the value of $f(x)$ is virtually indistinguishable from the value of $5x^4$. Thus, the value (and behavior) of $f(x)$ can be well-estimated by $5x^4$ as $x \to \pm\infty$.

6. For each polynomial function, identify the leading term.
   *a. $h(x) = -4x^3 + x^2 - 900x + 5$ *b. $f(x) = 9x^2 - 4x - 2x^4$ c. $g(x) = -7(x-5)(x^2+1)$

Since a polynomial function behaves like its leading term we can use notation to communicate this. Returning to $f(x) = 5x^4 + 7x^3 + 5x + 100$, we might write "As $x \to \pm\infty$, $f(x) \to 5x^4$." The arrow is read as "approaches" or "tends to".

However, we also know (or can easily determine) the end behavior of $y = 5x^4$. As $x \to \infty$, $5x^4 \to \infty$ and as $x \to -\infty$, $5x^4 \to \infty$. Thus, this is also the end behavior of $f(x)$. So we can say two things.

- As $x \to \pm\infty$, $f(x) \to 5x^4$. Also,
- As $x \to \infty$, $f(x) \to \infty$ and as $x \to -\infty$, $f(x) \to \infty$.

7. Using the same functions from Exercise #6, complete the following statements by first filling in the variable expression that well-estimates the function values and then filling in the end behavior in terms of increasing or decreasing without bound. *Part (a) is done for you.*
   a. $h(x) = -4x^3 + x^2 - 900x + 5$ *b. $f(x) = 9x^2 - 4x - 2x^4$ c. $g(x) = -7(x-5)(x^2+1)$

   As $x \to \pm\infty$, $h(x) \to -4x^3$.   As $x \to \pm\infty$, $f(x) \to$ _____.   As $x \to \pm\infty$, $g(x) \to$ _____.

   As $x \to \infty$, $h(x) \to -\infty$.   As $x \to \infty$, $f(x) \to$ _____.   As $x \to \infty$, $g(x) \to$ _____.

   As $x \to -\infty$, $h(x) \to \infty$.   As $x \to -\infty$, $f(x) \to$ _____.   As $x \to -\infty$, $g(x) \to$ _____.

---

8. Describe the end behavior for each of the following polynomial functions without graphing. Then check your work by graphing the functions with a graphing calculator or graphing software. *Remember to consider the role of the leading coefficient when determining a function's end behavior.*

    a. $f(x) = -x^3 + 4x^2 - 8$ b. $g(x) = 5x - x^4 + 3x^5$ c. $h(x) = -2(x+1)(x-3)$

9. Answer the following questions given the graph of *g*.

    a. What are the roots of *g*?

    b. Evaluate $g(0)$.

    c. On what interval(s) of the domain is the function increasing?

    d. On what interval(s) of the domain is the function decreasing?

    e. On what interval(s) of the domain does the function have positive concavity?

    f. On what interval(s) of the domain does the function have negative concavity?

    g. Estimate the location of any inflection points.

    h. Describe the end behavior of the function.

    i. When written as a formula, is the largest exponent an even number or an odd number? How do you know?

I. THE BOTTLE PROBLEM – MODELING CO-VARYING RELATIONSHIPS (TEXT: S2, 3, 4, 5)

1. a. A spherical fish bowl is filling with water. Describe how the height of the
       water and volume of water in the bowl vary together for equal volumes of
       water added to the bowl.
    b. Construct a graph that represents the height of the water in the bowl as a
       function of the volume of water in the bowl.

Spherical
Fish Bowl

2. The definition of an ***increasing function*** f follows: A function f is said to be increasing if
    $f(x_1) < f(x_2)$ whenever $x_1 < x_2$.
    a. Explain what this means in your own words. Use specific examples if it helps you.
    b. Given that g is a function defining the height of water in some bottle as a function of the volume
       of the water in the bottle, v, complete the following.
       i.   Why must g be an increasing function? (As v increases, why must $g(v)$ always increase?)
       ii. Explain the meaning of the statement $g(v_1) < g(v_2)$ whenever $v_1 < v_2$ in this context.
    b. Use mathematical symbols to convey that another function w is <u>decreasing</u> for all values of x.
       How does $w(x)$ change as x increases?

3. You are given four graphs representing the water height in a bottle as a function of the water volume.
    i.   Describe how the height of the water and volume of water in the bottle vary together for equal
        amounts of volume of water added to the bottle.
    ii. Construct a careful sketch of the bottle. Include landmarks on both the bottle and graph to show
        points where the function behavior changes in important ways.

    a.

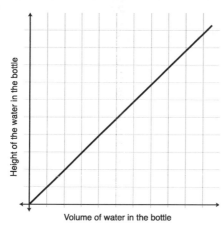

    b.

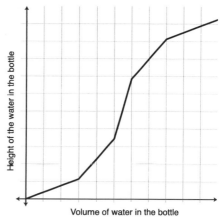

    c.

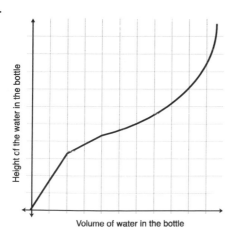

    d.

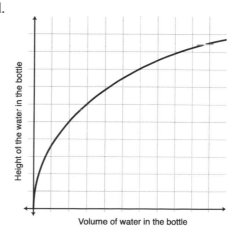

4. For each of the scenarios below, construct a graph of the height of the water in the bottle as a function of the volume of water in the bottle. Then, make an illustration of a bottle that would produce a height-volume graph with that general behavior.
    a. Scenario #1: A bottle in which the height of the water in the bottle increases at a constant rate with respect to the volume of water in the bottle.
    b. Scenario #2: A bottle in which the height of the water in the bottle increases at a constant rate with respect to the volume of water in the bottle, then switches so that equal changes in volume lead to smaller and smaller increases in the height of water in the bottle.
    c. Scenario #3: A bottle in which the height of the water in the bottle increases by larger and larger amounts for equal increases in volume, then increases at a constant rate with respect to the volume of water in the bottle, then increases by smaller and smaller amounts for equal increases in the height of water in the bottle.

5. As a runner is moving around a quarter mile track a radar gun detects the direct distance of the runner from the starting line.
    a. Construct a graph that represents the direct distance (in yards) of the runner from the starting line in terms of the total distance (in yards) the runner has traveled around the track.
    b. At approximately what point(s) around the quarter mile track is the direct distance of the runner from the starting line at its maximum value?
    c. Provide a written description of how the direct distance (in yards) of the runner from the starting line varies with the distance (in yards) the runner has traveled around the track.

6. Watch the video titled "Exploring Co-Varying Quantities" and answer the following questions. (The video can be found in the online textbook, Module 5 p. 7)
    a. What conditions are necessary for there to be a linear relationship between the height of the water and the volume of water in the bottle?
    b. For a cylindrical bottle with shoulders (like the one considered in the video) how does the height of the water in the bottle co-vary with the volume of water in the bottle?
    c. What do sharp corners on a height-volume graph represent?
    d. What is a technique that can be used when analyzing the co-variation of two quantities?

7. Expand the following expressions as much as possible. *Combine like terms to simplify your answer.*
    a. $(x-3)(x+4)$ b. $-(x-7)^2$ c. $(x-4)^2(x-1)$
    d. $3(x-2)^2$ e. $-(3x-2)(x+4)(2x-1)$

II. AVERAGE RATE OF CHANGE AND CONCAVITY (TEXT: S4, 5)

8. For the following functions determine if the average rates of change of the water height with respect to the water volume over successive equal-sized intervals is increasing, decreasing, or constant.

a.
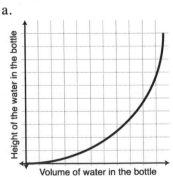

b.
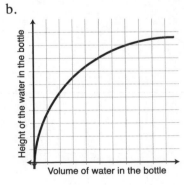

c.
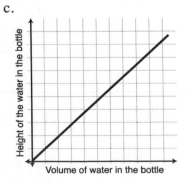

9. When graphing the height of the water in a bottle with respect to the volume of water in the bottle, ***inflection points*** on the height–volume graph correspond to where the bottle changes from getting narrower to getting wider (or vice-versa).

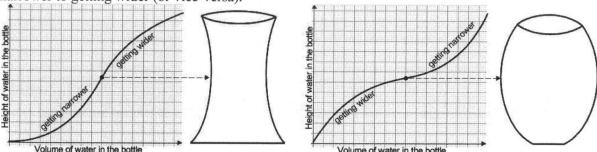

   a. Explain what the inflection point on the height-volume graph for the bottle on the left conveys about the changes in the water height and water volume as the bottle is being filled with water.
   b. Explain what the inflection point on the height-volume graph for the bottle on the right conveys about the changes in the water height and water volume as the bottle is being filled with water.

10. The values in the table below represent the distance (measured in yards) of a dog from a park entrance as a function of the number of seconds since the dog entered the park.

Change in the number of seconds since the dog entered the park	The number of seconds since the dog entered the park	The distance (in yards) of a dog from the park entrance	Change in the distance (in yards) of a dog from the park entrance	Average rate of change of the dog's distance with respect to time
	0	0		
	3	26.25		
	3.2	30.976		
	4.5	77.344		
	7	271.25		

   a. Complete the table of values.
   b. Describe what the average rates of change tell you about how the distance of the dog from the park entrance is changing over the time interval from $t = 0$ to $t = 7$ seconds.

11. For each of the following functions, estimate the intervals on which:
   i.   the function values are increasing
   ii. the function values are decreasing
   iii. the function has positive concavity
   iv. the function has negative concavity

   a.

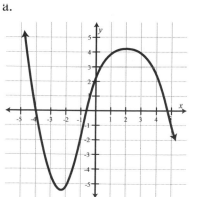

   b.

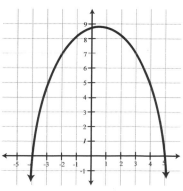

   c.

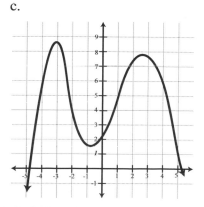

12. Using the functions in Exercise #11, do the following for each of the parts (a) through (c).
    i. Copy the graph.
    ii. Plot the points $(-2, f(-2))$ and $(3, f(3))$ on the graph.
    iii. Draw a line passing through the two points.
    iv. Find the constant rate of change of the line you drew. *You may have to estimate the function output values.*

Use the following context for Exercises #13-14. Fire pits are frequently used to cook meat. A fire is built to heat the rocks in the bottom of the pit. Once the pit reaches the desired temperature the meat is then put in the pit and covered.

13. The table below represents the air temperature (in degrees Fahrenheit) of a pit in terms of the number of hours that have elapsed since the meat was added to the pit.

Number of hours since meat was added to the pit	Air temperature (in degrees Fahrenheit) of the pit
0	575
2	400
4	320
6	275

   a. Using the table of values, do the following.
      i. Sketch a graph of the air temperature (in degrees Fahrenheit) of the pit in terms of the number of hours since the meat was added to the pit. Be sure to label your axes.
      ii. Represent the changes in the air temperature of the pit from 0 hours to 2 hours; from 2 hours to 4 hours; and from 4 hours to 6 hours.
      iii. What do you notice about the change in the air temperature of the pit for successive equal change is the number of hours since the meat was added to the pit?
   b. Let t represent the number of hours since the meat was added to the pit. Determine the average rate of change of the air temperature of the pit on the following time intervals and then describe how to interpret these values.
      i. $0 \le t \le 2$ ii. $2 \le t \le 4$ iii. $4 \le t \le 6$
   c. Does this function have positive concavity, negative concavity, or some combination of both on the interval $0 < t < 6$?

14. The table below represents the internal temperature (in degrees Fahrenheit) of the meat in terms of the number of hours that have elapsed since the meat was added to the pit.

Number of hours since meat was added to the pit	Internal temperature (in degrees Fahrenheit) of the meat
0	45
1	80
3	135
10	200

   Let t represent the number of hours since the meat was added to the pit. Determine the average rate of change of the internal temperature of the meat over the following time intervals and then describe how to interpret these values.
   a. $0 \le t \le 1$ b. $1 \le t \le 3$ c. $3 \le t \le 10$

15. $f(t) = 1.5^t$, with $d = f(t)$, represents a car's distance d (measured in feet) from a stop sign in terms of the number of seconds t since the car started to move away from the stop sign.
   a. Sketch a graph of this relationship. Be sure to label your axes.
   b. Determine the average rate of change of the distance of the car from the stop sign on the following time intervals, then explain how to interpret each value.
      i. $0 \le t \le 2$ ii. $2 \le t \le 4$ iii. $4 \le t \le 6$ iv. $1 \le t \le 5$
   c. Does this function have positive concavity, negative concavity, or some combination of both on the interval $0 < t < 6$?

16. Watch the video titled "Co-Varying Quantities and Changing Rate of Change of Non-Linear Polynomial Functions" and answer the following questions. (The video can be found in the online textbook, Module 5 p. 11)
    a. Provide an example of a context in which the values of the output quantity decrease and the average rates of change of the output quantity with respect to the input quantity are increasing as the value of the input quantity increases.
    b. Provide an example of a context in which the values of the output quantity decrease and the average rates of change of the output quantity with respect to the input quantity are decreasing as the value of the input quantity increases.
    c. For the values of r, a, n, and c given in the video is $r < n$ or is $n < r$? Explain your reasoning.
    d. Is -4 greater than or less than -30? Explain.

17. The given graph represents Sally's distance from her house in terms of the number of minutes t since Sally started walking.
    a. Interpret the meaning of the point (20, 1950).
    b. Represent the changes in Sally's distance from her house (in feet) from 0 minutes to 10 minutes; from 10 minutes to 20 minutes; and from 20 minutes to 30 minutes.
    c. What do you notice about the change in Sally's distance from her house for successive equal changes in the number of minutes since Sally started walking?
    d. Describe what these successive changes in Sally's distance from her house for equal change in the number of minutes since Sally started walking tell you about how Sally's distance from her house is changing over the time interval from $t = 0$ to $t = 30$ minutes.
    e. Describe how Sally's speed (in yards per minute) is changing as the number minutes since Sally started walking increases from 0 to 30 minutes.

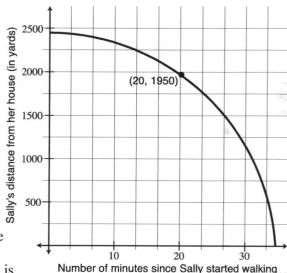

III. TRANSFORMATIONS OF POLYNOMIAL FUNCTIONS (TEXT: S9)

18. The graph of a polynomial function f is given. Sketch a graph of each the following functions.
    a. $g(x) = -f(x)$
    b. $h(x) = f(-x)$
    c. $k(x) = f(x-2)$
    d. $p(x) = f(x) - 2$

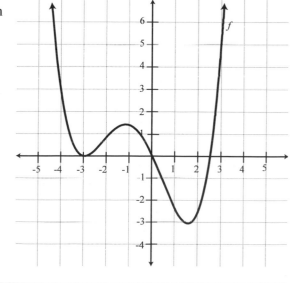

19. A coffee shop models its weekly profit P (measured in dollars) as a function of the price x (measured in dollars per cup) using $f(x) = -2000x^2 + 8000x - 2000$ with $P = f(x)$.

    a. A second coffee shop models its weekly profit as a function of the price it charges per cup using function g where $g(x) = f(x-1)$. Discuss the relationship between the inputs and outputs of functions g and f.

    b. A third coffee shop models its weekly profit as a function of the price it charges per cup using function h where $h(x) = f(x) - 150$. Discuss the relationship between the inputs and outputs of functions h and f.

    c. Do either g or h have the same maximum profits when compared to f? Explain.

20. The graphs of polynomial functions f and g are given below.
    a. How do the input and output pairs for the two functions compare?
    b. Express f in terms of g.
    c. Express g in terms of function f.

21. The graphs of polynomial functions f and g are given.
    a. How do the input and output pairs for the two functions compare?
    b. Express f in terms of g.
    c. Express g in terms of function f.
    d. How does the average rate of change for each function compare on the interval $-1 < x < 3$?

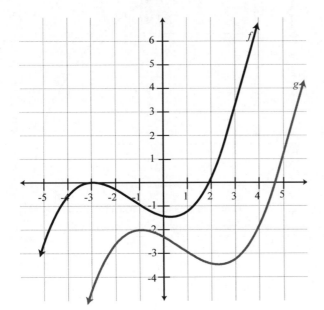

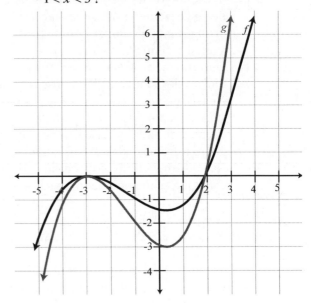

22. Each summer Primo Pizza and Pizza Supreme compete to see who has the larger summer profit. Let f be the function that determines Primo Pizza's profits in terms of the number of days since June 1. Let g be the function that determines Pizza Supreme's profits in terms of the number of days since June 1.

    a. If Primo Pizza's profits each day are two times as large as Pizza Supreme's profits,
        i. express the function f in terms of the function g.
        ii. express the function g in terms of the function f.

    b. If Primo Pizza's profits each day are two hundred dollars more than Pizza Supreme's profits,
        i. express the function f in terms of the function g.
        ii. express the function g in terms of the function f.

    c. If Primo Pizza's profits on a given day are always the same as Pizza Supreme's profits two days later,
        i. express the function f in terms of the function g.
        ii. express the function g in terms of the function f.

23. The function g is defined by $g(x) = 2f(x+3) - 4$. Explain how the behavior of function g compares to the behavior of function f.

24. Factor the following expressions as much as possible.
    a. $-8x^2 - 24x$ b. $x^2 + 6x - 16$ c. $2x^2 + 12x + 10$
    d. $2x^2 - 5x - 12$ e. $3x^2 - 27$ f. $6x^2 - x - 2$

IV. QUADRATIC FUNCTIONS (TEXT: S9)

25. Consider the functions $h(x) = x^2$ and $f(x) = 2^x$.
    a. How does the growth of the quadratic function h compare to the growth of the exponential function f?
    b. For large values of x, which of $f(x)$ or $h(x)$ will be larger?
    c. For what value(s) of x is the statement $f(x) = h(x)$ true?

26. Determine the roots of each quadratic function.
    a. $f(x) = (3x - 2)(x + 4)$ b. $g(x) = x^2 - 9$ c. $h(x) = 2x^2 + 5x + 3$ d. $k(x) = -3x^2 + 14x - 16$

27. Determine the vertex of each quadratic function.
    a. $g(x) = (3x - 7)(2x + 6)$ b. $j(x) = x^2 - 16$ c. $p(x) = 4x^2 + 8x + 3$

28. Determine the roots and axis of symmetry of each quadratic function.
    a. $f(x) = x^2 - 6x + 9$ b. $g(x) = (x + 3)(-2x - 5)$ c. $h(x) = -x^2 + 7x - 6$

29. Let $f(x) = x^2$, $g(x) = (x - 4)^2$, $h(x) = (x - 4)^2 + 7$, and $p(x) = 3(x - 4)^2 + 7$.
    a. Determine the roots of each function.
    b. Determine the axis of symmetry for each function.
    c. Sketch a graph of each function.
    d. Compare the graphs (including roots, vertex, and output values) of f, h, g, and p.

30. Given a quadratic function g defined by $g(x) = ax^2 + bx + c$, and the solutions $x = 2 \pm 3.4$ to the quadratic equation $ax^2 + bx + c = 0$, determine the following (if possible).
    a. the roots of g b. the vertex of g

31. Given a quadratic function g defined by $g(x) = ax^2 + bx + c$, and the solutions $x = -4.7 + 5$ to the quadratic equation $ax^2 + bx + c = 0$, determine the following (if possible).
    a. the roots of g b. the vertex of g

32. The function f represents the height of a ball h above the ground (measured in feet) in terms of the amount of time t (measured in seconds) since the ball was thrown upward from a bridge. Then $f(t) = -16t^2 + 48t + 120$ with $h - f(t)$.
    a. Approximately how high is the bridge above the ground? Justify your answer.
    b. When does the ball hit the ground? Justify your answer.
    c. Construct a rough sketch of the graph of the function f that relates the height of the ball above the ground and the amount of time since the ball was thrown from the bridge.

Exercise continues on the next page.

d. After how many seconds does the ball reach its maximum height above the ground? What is the maximum height above the ground reached by the ball? On the graph you drew in part (c) illustrate the point that represents the ball's maximum height above the ground.

e. Pick two points on the graph and determine the function's average rate of change over the interval between them. Explain how to interpret this value.

33. A penny is thrown from the top of a 30.48-meter building and hits the ground 3.45 seconds after it was thrown. The penny reached its maximum height above the ground 0.823 seconds after it was thrown.

a. Define a quadratic function h that expresses the height of the penny above the ground (measured in meters) as a function of the elapsed time (measured in seconds) since the penny was thrown. (*Hint: Determine the zeros of the quadratic function, and then use the height of the building to find the value of the leading coefficient.*)

b. What is the maximum height of the penny above the ground?

34. A rock is thrown upward from a bridge that is 20 feet above a road. The rock reaches its maximum height above the road 0.91 seconds after it is thrown and contacts the road 2.35 seconds after it was thrown. Use your knowledge of the symmetry of a parabola and the given information to develop a quadratic function that relates the time since the rock was thrown to the height of the rock above the bridge.

35. A coffee shop finds that its weekly profit P (measured in dollars) is determined by the price x (in dollars) that the coffee shop charges for a cup of coffee. The profit P can be determined by the function $f(t) = -4000x^2 + 12000x - 4000$ with $P = f(t)$.

a. Determine the maximum weekly profit and the price for a cup of coffee that produces the maximum profit.

b. If $g(x) = f(x-2)$, describe how the inputs and outputs of g and f are related. Does g have the same maximum profit as f? Explain.

c. If $h(x) = f(x) - 2$, describe how the inputs and outputs of h and f are related. Does h have the same maximum profit as f? Explain.

d. If $p(x) = f(x) + 60$, describe how the inputs and outputs of p and f are related.

36. Determine if each of the following are true or false. Provide a brief justification for your answers.

a. A parabola must have either zero or two x-intercepts.

b. If $f(x) = ax^2 + bx + c$ with $y = f(x)$, the vertical intercept of function is $y = c$.

c. The vertex of a parabola is the point where the function values change from increasing to decreasing, or decreasing to increasing.

d. The zeros of a quadratic function are the horizontal intercept(s) of the function's graph.

e. The domain of a quadratic function (ignoring contexts) includes all real numbers.

f. The range of a quadratic function (ignoring contexts) includes all real numbers.

g. If a quadratic function f has negative concavity on the interval $0 < x < 4$ then the values of $f(x)$ must decrease on this interval.

37. Use the quadratic formula to determine the roots of the following functions.

a. $f(x) = x^2 + 5x$ b. $g(x) = x^2 + 4x - 3$ c. $k(x) = 10x^2 - 8x - 21$ d. $m(x) = -3x^2 - 2 + 12x$

38. Given the quadratic functions in factored form, write them in an equivalent standard form. Then construct a graph of each function and label its roots and maximum or minimum value.

a. $h(x) = (2x)(3x - 4)$ b. $m(x) = (x-5)(2x+3)$ c. $f(x) = -(3x+3)(2x-4)$

39. Given the quadratic functions in standard form, write them in an equivalent factored form. Then construct a graph of each function and label its roots and maximum or minimum value.
    a. $p(x) = 2x^2 - 5x - 3$　　　b. $h(x) = 3x^2 + 11x - 4$　　　c. $k(x) = -x^2 + 9x - 14$

40. a. Define the formula for a quadratic function that has horizontal intercepts (roots) at $x = -6$ and $x = 2$ and a graph that passes through the point $(0, 5)$.
    b. What is the parabola's vertex?

41. a. Define the formula for a quadratic function that has horizontal intercepts (roots) at $x = 1$ and $x = 3$ and a graph that passes through the point $(0, -10)$.
    b. What is the parabola's vertex?

42. Use the quadratic formula to solve the following equations for x. (*Hint: First rewrite the equation so that it equals 0. This will allow you to utilize the quadratic formula.*)
    a. $-7x^2 + 13x = -6$　　b. $3x^2 - 4x - 4 = 12$　　c. $-6x^2 + 22 = 35x$　　d. $2x^2 + 5x = 12x - 2$

43. Determine the vertex of each quadratic function by completing the square. (*Note that the product of completing the square only changes the form of a quadratic function. The values represented by the function are unchanged.*)
    a. $f(x) = x^2 - 4x + 1$　　　b. $h(x) = x^2 + 8x - 7$　　　c. $k(x) = 2x^2 + 8x - 3$
    d. $p(x) = -2x^2 + 6x - 1$　　e. $m(x) = -x^2 + 14x + 3$　　f. $p(x) = 3x^2 + 2x + 1$

44. Given the quadratic functions in factored form, write them in equivalent standard form and then complete the square to find the function's maximum or minimum value. Finally construct a graph of each function and label its roots and maximum or minimum value.
    a. $f(x) = (2x + 1)(x - 5)$　　　　　b. $h(x) = (x - 7)(-3x + 4)$

45. Why are second order polynomials called quadratics? *You may conduct research on the Internet to determine the answer.*

46. Watch the video titled "Rock Throw Revisited: Creating the Formula" and answer the following questions. (*The video can be found in the online textbook, Module 5 p. 49.*)
    a. Explain how to determine the value of one root of a quadratic function given the other root and the vertex.
    b. Which of the following represents the roots of a function that models the rock throwing problem discussed in the video?
       i.　0.91 ± 1.44　　　　　ii.　2.35 ± 2　　　　　iii.　2.35 ± 0.53
    c. Why is the factored form of a function useful?
    d. Suppose the zeros of a quadratic function are $x = 2$ and $x = -0.35$ and the graph of the function passes through the point $(0, 2)$. Determine the function's formula.

V. Roots and End Behavior of Polynomial Functions (Text: S1, 6, 7, 8)

47. What is the general form of a polynomial function formula? Identify the constant term and leading coefficient.

48. Which of the following are polynomial functions. Justify your answer.
    a. $f(x) = 2^x$　　b. $g(x) = 5$　　c. $h(x) = \frac{5}{x} - 3x^2$　　d. $p(x) = 5x^4 + 3x^2 - 122$　　e. $r(x) = 5^{-2} - 3x$

49. Determine the roots of the following polynomial functions.
    a. $g(x) = 3x(2x - 4)(x + 2)^2$
    b. $h(x) = x^2(x - 4)^2(x^3 - 8)$
    c. $h(x) = x^3 + 6x^2 + 3x$
    d. $k(x) = 2x^2 - 5x - 3$

50. Determine the roots of the following polynomial functions.
    a. $f(x) = 4x^2 + 8x + 2$
    b. $g(x) = x^3 - x^2$
    c. $s(x) = -(2x + 4)(3x - 1)^2(x - 2)$
    d. $n(x) = 2x(x - 7)(3x + 4)$

51. Given the function $f(x) = (x - 3)(x + 1)(x - 2)$, complete the following.
    a. Determine the roots of f.
    b. As x increases without bound, describe the behavior of $f(x)$.
    c. As x decreases without bound, describe the behavior of $f(x)$.
    d. What is the behavior of f on the intervals $-1 < x < 2$ and $2 < x < 3$?
    e. Use your responses in parts (a) through (d) to sketch a graph of f.

52. For the following polynomial functions determine:
    i. the interval(s) on which the output of the function is positive.
    ii. the interval(s) on which the output of the function is negative.
    a. $f(x) = x(3x + 6)(x - 1)$
    b. $g(x) = 2x^3 - 4x^2 + 4x$
    c. $h(x) = 3x(2x - 5)(x + 1)$

53. Define three polynomial functions that have roots at $x = 2$, $x = 4$, and $x = -3$.

54. Define three polynomial functions that have roots at $x = 1$, $x = -1$, and $x = -5$.

55. Find the formula for a polynomial function that has horizontal intercepts (roots) at $x = 2$, $x = 5$, and $x = -4$ and passes through the point $(3, 6)$.

56. For each of the polynomial functions given, identify the leading term of the function and describe the end behavior of the function based on your analysis of the leading term.
    a. $f(x) = 12x^3 - 2x^5 + 3x - 2$
    b. $g(x) = 3x(x - 2)(x + 4)(-2x + 3)^2$
    c. $h(x) = 3x^3 - 4x + 8x^{10} - 3$
    d. $m(x) = -2(4 + x)(2x - 7)$

57. For each of the polynomial functions given, identify the leading term of the function and describe the end behavior of the function based on your analysis of the leading term.
    a. $f(x) = 3x(x - 7)(x + 2)(x - 4)$
    b. $h(x) = 2x^2(-x + 4)(x - 7)^2$
    c. $p(x) = 2x(3x - 7)(4x + 1)$
    d. $s(x) = 3x^2 + 4x - 7x^8 + 6x - 2$

58. For each of the polynomial functions given, identify the leading term of the function and describe the end behavior of the function based on your analysis of the leading term.
    a. $h(x) = 2x(x + 5)(x - 1)^2$
    b. $k(x) = -(x + 3)(x - 4)^2$
    c. $g(x) = x^6 - 7x^5 + 13x^4 + 7x^3 - 34x^2 + 4x + 24$
    d. $f(x) = x^3 - 4x^2 + x + 6$

59. Given the function $f(x) = x^2(3x-4)(2x+5)^3$, complete the following.
    a. Determine the roots of f.
    b. Describe whether the graph of f will cross through (change signs from positive to negative or vice versa) or only "bounce off" the horizontal axis at each of the roots.
    c. Evaluate $f(0)$.
    d. Examine the leading term of f to determine the end-behavior of the function.
    e. Determine on what interval(s) of the domain $f(x)$ is positive.
    f. Determine on what interval(s) of the domain $f(x)$ is negative.
    g. Use your responses in part (a) through (f) to sketch a graph of f.

60. Given the function $g(x) = x^3(2x-4)^2(3x-2)(-x+1)^2$, complete the following.
    a. Determine the roots of g.
    b. Describe whether the graph of g will cross through (change signs from positive to negative or vice versa) or only "bounce off" the horizontal axis at each of the roots.
    c. Evaluate $g(0)$.
    d. Examine the leading term of g to determine the end-behavior of the function.
    e. Determine on what interval(s) of the domain $g(x)$ is positive.
    f. Determine on what interval(s) of the domain $g(x)$ is negative.
    g. Use your responses in part (a) through (f) to sketch a graph of g.

Use the graph of f to complete Exercises #61-63.

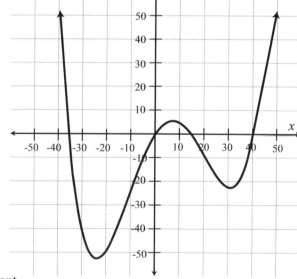

61. a. What is the domain of f?
    b. What is the range of f?
    c. Evaluate $f(0)$.
    d. What are the roots of f?

62. a. On what interval(s) is $f(x)$ increasing?
    b. On what interval(s) is $f(x)$ decreasing?
    c. On what interval(s) does f have positive concavity?
    d. On what interval(s) does f have negative concavity?

63. a. When written as a formula, is the largest exponent an even number or an odd number? Explain your reasoning.
    b. Describe the behavior of $f(x)$ as x increases without bound.
    c. Describe the behavior of $f(x)$ as x decreases without bound.
    d. Are there any roots with an even multiplicity (even exponent)?

Use the graph of *g* to complete Exercises #64-66.

64. a. What is the domain of *g*?
    b. What is the range of *g*?
    c. Evaluate $g(0)$.
    d. What are the roots of *g*?

65. a. On what interval(s) is $g(x)$ increasing?
    b. On what interval(s) is $g(x)$ decreasing?
    c. On what interval(s) does *g* have positive concavity?
    d. On what interval(s) does *g* have negative concavity?

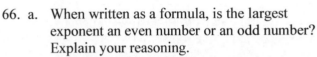

66. a. When written as a formula, is the largest exponent an even number or an odd number? Explain your reasoning.
    b. Describe the behavior of $g(x)$ as *x* increases without bound.
    c. Describe the behavior of $g(x)$ as *x* decreases without bound.
    d. Are there any roots with an even multiplicity (even exponent)?

67. Evaluate each of the following
    a. $f(x) = (8.5 - 2x)(11 - 2x)x$ when $x = 3.4$ b. $g(x) = 87.4 - 53x$ when $x = 1.2$
    c. $h(x) = 3x^3 + 2x - 7$ when $x = -3.2$

68. Determine if each of the following are true or false. Provide a brief justification for your answers.
    a. The process of finding zeros of a polynomial function is also referred to as finding the roots of the polynomial function.
    b. We can apply the zero-product property to determine the zeros of a polynomial function in factored form.
    c. The function used to model the box problem in Module 3 has two roots, at $x = 0$ and $x = 4.25$.
    d. The zeros of any function *f* represent the value(s) of *x* that when input into the function *f* return a value of 1 for the output $f(x)$.
    e. Polynomial functions are not always continuous.

69. Watch the video titled "End Behavior of Polynomial Functions" and answer the following questions. (*The video can be found in the online textbook, Module 5 p. 31.*)
    a. How do we represent the phrase "*x* increases without bound"?
    b. Why is it worthwhile to consider the relative magnitude of each term in a polynomial function?
    c. True or False: When we consider the end behavior of a function we look at how the output values change for input values close to zero. *If the statement is false, rewrite the statement so it is true.*
    d. Given $f(x) = x^3 - 2x^2 + 5x - 100$, what terms dominates the value of $f(x)$ as *x* increases without bound?

This investigation contains review and practice with important skills and procedures you may need in this module and future modules. Your instructor may assign this investigation as an introduction to the module or may ask you to complete select exercises "just in time" to help you when needed. Alternatively, you can complete these exercises on your own to help review important skills.

Simplifying Radicals
Use this section prior to the module or at any point as a spiraled review of skills.

One property of radicals is represented by the statement $\sqrt[n]{a \cdot b} = \sqrt[n]{a} \cdot \sqrt[n]{b}$. Basically, the nth root of a product is equivalent to finding the nth root of each factor and multiplying the results.

Three examples are shown below that demonstrate this property.

$$\sqrt{64} \qquad\qquad \sqrt{100} \qquad\qquad \sqrt[3]{216}$$
$$\sqrt{16 \cdot 4} \qquad\qquad \sqrt{25 \cdot 4} \qquad\qquad \sqrt[3]{27 \cdot 8}$$
$$\sqrt{16} \cdot \sqrt{4} \qquad\qquad \sqrt{25} \cdot \sqrt{4} \qquad\qquad \sqrt[3]{27} \cdot \sqrt[3]{8}$$
$$4 \cdot 2 \qquad\qquad 5 \cdot 2 \qquad\qquad 3 \cdot 2$$
$$8 \qquad\qquad 10 \qquad\qquad 6$$

These examples don't demonstrate WHY you would use this property (for example, we already knew that $\sqrt{64} = 8$). This property is most commonly used to rewrite irrational numbers where the number inside can be written as a product of a perfect root and some other number. See the examples below.

$$\sqrt{50} \qquad\qquad \sqrt{192} \qquad\qquad \sqrt[3]{56}$$
$$\sqrt{25 \cdot 2} \qquad\qquad \sqrt{64 \cdot 3} \qquad\qquad \sqrt[3]{8 \cdot 7}$$
$$\sqrt{25} \cdot \sqrt{2} \qquad\qquad \sqrt{64} \cdot \sqrt{3} \qquad\qquad \sqrt[3]{8} \cdot \sqrt[3]{7}$$
$$5\sqrt{2} \qquad\qquad 8\sqrt{3} \qquad\qquad 2\sqrt[3]{7}$$

Take a moment to verify that the expressions at the beginning and end of each process are equivalent. For example, verify that $\sqrt{192}$ and $8\sqrt{3}$ have the same decimal approximation.

In Exercises #1-10 rewrite each expression so that the number inside the radical is as small as possible. *For example, rewrite $\sqrt{18}$ as $2\sqrt{3}$.*

1. $\sqrt{75}$ 　　　 2. $\sqrt{28}$ 　　　 3. $\sqrt{48}$ 　　　 4. $\sqrt{108}$ 　　　 5. $\sqrt{45}$

6. $\sqrt{98}$ 　　　 7. $\sqrt[3]{54}$ 　　　 8. $\sqrt[3]{72}$ 　　　 9. $\sqrt[4]{48}$ 　　　 10. $\sqrt[4]{162}$

Rewriting Expressions
Use this section prior to the module or with/after Investigations 1 and 2.

In Exercises #11-16 use the distributive property to rewrite each expression.

11. $2x(x^2 - 6)$　　　　　　12. $-x(10 - x^2)$　　　　　　13. $4x(x - 3y)$

14. $x^3(2x^4 - \frac{1}{2}x)$　　　　　15. $-\frac{2}{3}x(6x + 10)$　　　　16. $-\frac{7}{2}xy(8xy^2 - x^5y^3)$

In Exercises #17-28 rewrite each expression in expanded form. *For example, write $(x + 2)(2x + 5)$ in the form* $2x^2 + 9x + 10$.

17. $(x + 4)(x + 5)$　　　18. $(x - 3)(x + 6)$　　　19. $(x + 7)(x - 7)$

20. $(x + 10)(x - 8)$　　　21. $(x - 3)(x - 6)$　　　22. $(x - 10)(x + 10)$

23. $(x - 1)(x - 4)$　　　24. $(x + 6)^2$　　　25. $(x - 5)^2$

26. $(2x + 7)(x + 2)$　　　27. $(4x - 5)(4x + 5)$　　　28. $(2x + 3)(3x - 2)$

In the next set of exercises we will factor expressions. A few things to keep in mind when you work on these. First, remember that it can sometimes be helpful to identify common factors. For example, factoring $2x^3 - 16x^2 + 30x$ is easier when we recognize that $2x$ is a common factor of all three terms.

$$2x^3 - 16x^2 + 30x$$
$$2x(x^2 - 8x + 15)$$
$$2x(x - 3)(x - 5)$$

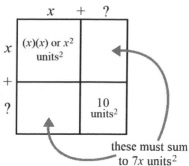

Second, recall the "area model" introduced in an earlier module. This model is often quite helpful for students who are trying to factor trinomial expressions like $x^2 + 7x + 10$. In this case the factored form is $(x + 2)(x + 5)$.

Third, recall that sometimes a "middle term" can have a value of 0. For example, if you want to factor the expression $x^2 - 25$ you might notice that there is no "x" term (only a term with x^2). However, without changing its value, we can rewrite the expression $x^2 - 25$ as $x^2 + 0 - 25$, or even better as $x^2 + 0x - 25$. The following examples demonstrate this idea.

$$x^2 - 25 \qquad\qquad x^2 - 64 \qquad\qquad 4x^2 - 9y^2$$
$$x^2 + 0x - 25 \qquad\qquad x^2 + 0x - 64 \qquad\qquad 4x^2 + 0xy - 9y^2$$
$$(x + 5)(x - 5) \qquad\qquad (x + 8)(x - 8) \qquad\qquad (2x + 3y)(2x - 3y)$$

29. a. Expand $(x - 6)(x + 6)$ to show that it is equivalent to $x^2 - 36$.

    b. Expand $(2x - 5)(2x + 5)$ to show that it is equivalent to $4x^2 - 25$.

In Exercises #30-41 factor each expression. *Use the tips just discussed to help you.*

30. $x^2 + 16x + 60$ 31. $x^2 - 5x - 24$ 32. $x^2 - 100$

33. $5x^3 + 30x^2 + 40x$ 34. $x^2 - 144$ 35. $2x^2 - x - 28$

See instructions on the previous page for Exercises #36-41.

36. $4x^2 - 49$

37. $x^2 - 2xy + y^2$

38. $2x^3 + 5x^2 - 12x$

39. $16x^2 - 36$

40. $3a^2 + 6ab + 3b^2$

41. $6x^3 + 21x^2 + 15x$

Simplifying Rational Expressions

Use this section prior to the module or with/after Investigations 2 and 3.

Sometimes it's possible to factor out the same number from the numerator and denominator of a rational expression. This makes it possible to simplify the expression. For example, the numerator and denominator of $\dfrac{5x+35}{10}$ both have 5 as a common factor.

$$\frac{5x+35}{10}$$

$$\frac{5(x+7)}{5(2)}$$

Since $\frac{a}{b} \cdot \frac{c}{d} = \frac{a \cdot c}{b \cdot d}$ (which is true in "both directions"), we can rewrite this as follows.

$$\frac{5}{5} \cdot \frac{x+7}{2}$$

$$1 \cdot \frac{x+7}{2}$$

$$\frac{x+7}{2}$$

Thus, $\dfrac{5x+35}{10}$ can be simplified to $\dfrac{x+7}{2}$. They are equivalent rational expressions. We demonstrate two more examples below.

$$\frac{12x-15}{3}$$
$$\frac{3(4x-5)}{3(1)}$$
$$\frac{3}{3} \cdot \frac{4x-5}{1}$$
$$1 \cdot \frac{4x-5}{1}$$
$$4x-5$$

$$\frac{14x+28}{21}$$
$$\frac{7(2x+4)}{7(3)}$$
$$\frac{7}{7} \cdot \frac{2x+4}{3}$$
$$1 \cdot \frac{2x+4}{3}$$
$$\frac{2x+4}{3}$$

In Exercises #42-47 simplify the rational expression if possible. If you can't simplify the expression, write "does not simplify".

42. $\dfrac{10x-20}{10}$

43. $\dfrac{8x-12}{9}$

44. $\dfrac{21x+3y}{3}$

45. $\dfrac{8x+40}{32}$

46. $\dfrac{35x-5y}{10}$

47. $\dfrac{18x-72}{45}$

It's also possible to simplify rational expressions when the numerator and denominator have the same variable factors.

$$\frac{2x^2+10x}{18x}$$

$$\frac{2x(x+5)}{2x(9)}$$

$$\frac{2x}{2x}\cdot\frac{x+5}{9}$$

$$1\cdot\frac{x+5}{9}$$

$$\frac{x+5}{9}$$

Continued on the next page.

HOWEVER, it's important to specify that $\dfrac{2x^2+10x}{18x}$ and $\dfrac{x+5}{9}$ are equivalent only as long as $x \neq 0$.

When $x=0$ the original expression $\dfrac{2x^2+10x}{18x}$ is undefined. So we say that $\dfrac{2x^2+10x}{18x} = \dfrac{x+5}{9}$ when $x \neq 0$. Two more examples follow.

$$\frac{x^2+6x+5}{x+5}$$
$$\frac{(x+5)(x+1)}{(x+5)(1)}$$
$$\frac{x+5}{x+5} \cdot \frac{x+1}{1}$$
$$1 \cdot \frac{x+1}{1}$$
$$x+1 \quad \text{if } x \neq -5$$

$$\frac{x^2-7x+12}{x^2-2x-8}$$
$$\frac{(x-4)(x-3)}{(x-4)(x+2)}$$
$$\frac{x-4}{x-4} \cdot \frac{x-3}{x+2}$$
$$1 \cdot \frac{x-3}{x+2}$$
$$\frac{x-3}{x+2} \quad \text{if } x \neq 4$$

Notice how we excluded $x=-5$ in the first example because it made the original expression undefined (but not the simplified form). We excluded $x=4$ in the second example for the same reason.

In Exercises #48-53 simplify the rational expression if possible. If you simplify the rational expression, be sure to list x-values we must restrict. If you can't simplify the expression, write "does not simplify".

48. $\dfrac{x^2+5x-14}{x-2}$

49. $\dfrac{x^4+7x^2y}{3x^2}$

50. $\dfrac{x^2-8x-33}{x^2+5x+6}$

51. $\dfrac{x^2-7x+2}{x}$

52. $\dfrac{x+1}{3x^2-x-4}$

53. $\dfrac{x^3}{x^3-2x^2+5x}$

*1. The National Center for Education Statistics (nces.ed.gov) keeps careful records of the number of degrees awarded in the United States. The given table shows the number of PhDs awarded to men and women in the U.S. over time.

- Let $f(t)$ model the number of PhDs awarded to men in terms of the year, t.
- Let $g(t)$ model the number of PhDs awarded to women in terms of the year, t.

a. What is the value of $f(1940)$? $g(1890)$?

year, t	# of PhDs awarded to men in the U.S., $f(t)$	# of PhDs awarded to women in the U.S., $g(t)$	
1880	51	3	
1890	147	2	
1900	359	23	
1920	522	93	
1940	2,861	429	
1960	8,801	1,028	
1980	69,526	26,105	
2000	64,171	55,414	
2010	76,605	81,953	

b. Let h be the function that inputs the year, t, and outputs the value of the ratio $\frac{f(t)}{g(t)}$. That is, $h(t) = \frac{f(t)}{g(t)}$. What is the value of $h(1880)$ and what does it represent in this context?

c. As t varies from 1890 to 1900, does the value of $h(t)$ increase or decrease? Why?

d. Using the given axes, label the axes based on the input and output quantities for h and then plot the points $\left(1880, h(1880)\right)$, $\left(1890, h(1890)\right)$, and $\left(1900, h(1900)\right)$. *Note: We have drawn the horizontal axis as a "broken" axis – you can allow the first tick mark to represent 1870.*

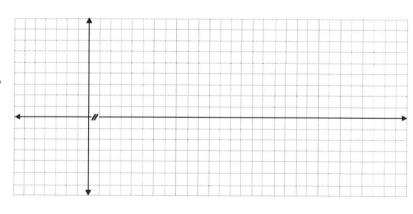

e. What must be true in this context if $h(t) = 3$ for some value of t? If $h(t) = 1$?

f. Is it possible for the output of h to be less than 1 in this context? Explain.

g. Is it possible for the output of h to be negative in this context? Explain.

h. Must there be a maximum value for $h(t)$ in this context? Explain your thinking.

i. Use the empty column on the previous page to record values of $h(t)$ and then plot each of the entries in the column as points on your graph. With a partner or as a class, describe the general behavior of h and what that behavior indicates about this context.

Function Outputs as Ratio Values

In Exercise #1 the outputs of function h represented the value of a ratio of two other varying quantities. Function h's output didn't tell us how many PhDs were awarded to men or women – it told us about the <u>relative size</u> of these quantities.

This presents interesting scenarios. For example, it's possible for the number of PhDs awarded to men and women to both increase but for the value of the ratio to decrease. When working with a function whose output is a ratio of two values we <u>must</u> pay careful attention to *how* the two quantities change and how this affects their relative size if we want to be able to understand the meaning of the output values and understand the function's behavior.

---

*2. Consider the function $f(x) = \frac{2}{x}$. What happens to the value of the function when x gets close to 0? Use the given tables to help you formulate your response.

x	$f(x) = \frac{2}{x}$
−2	
−1	
−0.1	
−0.01	
−0.001	

x	$f(x) = \frac{2}{x}$
0.001	
0.01	
0.1	
1	
2	

"Approaches" Notation

In Exercise #2 we thought about the value of x getting close to $x = 0$ from values less than 0 and from values greater than 0.

When x approaches 0 from values less than 0 (increasing towards 0) we say that "x approaches 0 from the left" and write this in notation as $x \to 0^-$ (*because the direction comes from the negative end of the number line*).

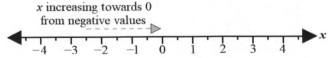

When x approaches 0 from values greater than 0 (decreasing towards 0) we say that "x approaches 0 from the right" and write this in notation as $x \to 0^+$ (*because the direction comes from the positive end of the number line*).

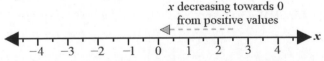

---

3. The following sequences represent varying values of x. Use "approaches" notation to describe the trend you observe.

   *a. 6, 5, 4.6, 4.55, 4.51, 4.5001, 4.500001, ...

   b. 9, 9.9, 9.99, 9.999, 9.9999, ...

   *c. $-6, -5.5, -5.1, -5.01, -5.001, ...$

   d. $-1, -1.5, -1.9, -1.99, -1.999, ...$

*4. Determine whether the following sequences of numbers are increasing or decreasing (*Note: A sequence of all negative numbers is increasing if successive numbers are getting closer to 0.*)

   a. $\frac{1}{100}, \frac{1}{200}, \frac{1}{320}, \frac{1}{450}, \frac{1}{1089}$

   b. $\frac{2}{100}, \frac{5}{40}, \frac{10}{35}, \frac{15}{9}, \frac{50}{6}$

   c. $\frac{100}{2}, \frac{40}{5}, \frac{35}{10}, \frac{9}{15}, \frac{6}{50}$

   d. $\frac{-1}{100}, \frac{-1}{200}, \frac{-1}{320}, \frac{-1}{450}, \frac{-1}{1089}$

*5. If $f(x) = \frac{1}{x-8}$, complete the following.

   a. Complete the following table of values. *Pay attention the value of the numerator and denominator as you perform your calculations.*

x	$f(x)$
7	
7.5	
7.9	
7.99	
7.999	

x	$f(x)$
8.001	
8.01	
8.1	
8.5	
9	

   b. What do you notice as x approaches 8 from the left (from values less than 8, written $x \to 8^-$)? Why does this happen?

   c. What do you notice as x approaches 8 from the right (from values greater than 8, written $x \to 8^+$)? Why does this happen?

   d. Using a calculator, graph f. Discuss with a group or as a class how the graph supports your answers in parts (b) and (c).

> ### Vertical Asymptotes
>
> A ***vertical asymptote*** occurs at a real number $x = a$ if
> - as $x \to a^-$ (as x approaches a from the left, or from values less than a) the value of $f(x)$ increases or decreases without bound
>
>   **and**
>
> - as $x \to a^+$ (as x approaches a from the right, or from values greater than a) the value of $f(x)$ increases or decreases without bound

Note that the examples in Exercises #2 and #5 were functions whose output values represented a ratio of two polynomial functions. Mathematicians call these type of functions ***rational functions***.

> ### Rational Functions
>
> If p and r are polynomial functions, then a function whose output values represent a ratio of the values of these functions (such as $f(x) = \frac{p(x)}{r(x)}$) is called a ***rational function.***

6. Suppose $f(x) = \frac{p(x)}{r(x)}$ where p and r are polynomial functions.

   a. For what value(s) of x is $f(x)$ undefined?

   b. For what value(s) of x does $f(x) = 0$?

*7. Five gallons of liquid flavoring are poured into a large vat. Water will be added to the vat and mixed with the flavoring to produce a drink that will be bottled and sold.

   a. Suppose water is added until the total mixture is 7 gallons. What is the ratio of flavoring to water? What does this ratio represent? (Hint: *How many times as large…*)

   b. Suppose water is added until the total mixture is 18 gallons. What is the ratio of flavoring to water?

   c. Define a function f that determines the ratio, $R,$ of flavoring to water in the mixture in terms of the total volume of the mixture x (in gallons). What is the practical domain of this function?

   d. What does it mean to say "as $x \to 5^+$ "?

   e. What happens to the value of $f(x)$ as $x \to 5^+$? Why does this make sense?

8. The graphs of $y = p(x)$ and $y = r(x)$ are given below. The rational function h is defined by $h(x) = \frac{p(x)}{r(x)}$.
   Let's explore the behavior of h by thinking about the behavior of p and r.

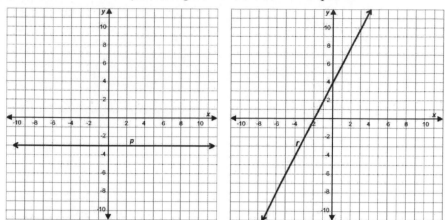

a. As x increases from -10 to -2, describe whether the value of each function is 1) positive or negative and 2) constant, increasing or decreasing.
   i. The value of $p(x)$ is...   ii. The value of $r(x)$ is...   iii. The value of $h(x)$ is...

b. Evaluate $h(-2)$.

c. As x increases from -2 to 0, describe whether the value of each function is 1) positive or negative and 2) constant, increasing or decreasing.
   i. The value of $p(x)$ is...   ii. The value of $r(x)$ is...   iii. The value of $h(x)$ is...

d. Evaluate $h(0)$.

e. As x increases from $x = 0$ to $x = 5$, describe the behavior of h.

f. Sketch a graph of the function h.

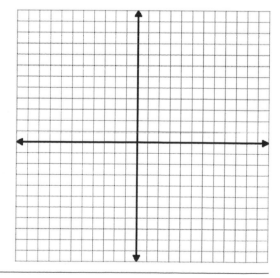

9. For each of the following functions, predict where a vertical asymptote will exist. Then test your prediction using a graphing calculator.

a. $f(x) = \dfrac{8}{x+1}$

b. $g(x) = \dfrac{x}{2x-8}$

*c. $h(x) = \dfrac{6x^2 - 24}{2(x-3)(x+1)}$

d. $m(x) = \dfrac{x+3}{x^2 + 4x - 5}$

*e. $n(x) = \dfrac{x^5 + 1}{x^2 + 1}$

*f. $c(x) = \dfrac{2(x+6)}{x+6}$

10. Write an explanation describing how to determine the vertical asymptotes of a rational function and *why* this approach works.

*1. Interpret the meaning of each of the following.

   a. $x \to 4^+$ b. $x \to 6^-$ c. $x \to \infty$ d. $f(x) \to -\infty$

In Module 5 we explored the long-term behavior (or end behavior) of polynomial functions. We know that as x increases or decreases without bound the leading term is the largest magnitude term in the polynomial. Therefore, the behavior of the function is well-estimated by the behavior of the leading term.

If $f(x) = x^2 - 10x + 5$, then as $x \to \pm\infty$ (as x increases or decreases without bound), f behaves indistinguishably from $y = x^2$. Thus, we might say the following.

$$\text{As } x \to \pm\infty, \ f(x) \to x^2.$$

How might this help us understand the long-term behavior of rational functions?

*2. Given $f(x) = \frac{2x^2 - 6x + 1}{x - 5}$ (where $2x^2 - 6x + 1$ and $x - 5$ are polynomials), complete the following.

   a. As x increases or decreases without bound (as $x \to \pm\infty$), what expression can we use to estimate the behavior of $2x^2 - 6x + 1$?

   b. As $x \to \pm\infty$, what expression can we use to estimate the behavior of $x - 5$?

   c. As $x \to \pm\infty$, what expression do you think we can use to estimate the behavior of $f(x) = \frac{2x^2 - 6x + 1}{x - 5}$?
      [*Simplify the expression if possible.*]

   d. Using a graphing calculator or graphing software, graph function f along with $y = \#\#$ on the same set of axes, where "##" is the expression you determined in part (c).
      i. When x is relatively close to 0 (from $x = -10$ to $x = 10$, for example), do the two functions behave similarly?

      ii. Zoom out on the graph. As you zoom out, what do you notice?

      iii. What do your answers to parts (i) and (ii) tell us about the rational function f?

3. For each of the following rational functions, create an expression that estimates the function's end behavior (its behavior as $x \to \pm\infty$). Simplify your answer and then use a graphing calculator to check your work (see Exercise #2 part (d)).

   a. $f(x) = \frac{4x + 8}{2x - 9}$ b. $g(x) = \frac{x^3 + x - 20}{x + 4}$ c. $h(x) = \frac{x^2 - 6x + 5}{x^3 + 2}$

4. A small airport is considering selling jet fuel at its airport. There is an initial investment of $3,200,000 to install tanks for the fuel. The price the airport pays for fuel is $3.90 per gallon. The function $f(x) = 3,200,000 + 3.90x$ models the total cost (in dollars) for the airport to purchase x gallons of fuel (including the startup cost and per-gallon cost).

   a. The average cost per gallon of fuel can be modeled by the function g where
   $g(x) = \dfrac{f(x)}{x} = \dfrac{3,200,000 + 3.90x}{x}$. Determine the values of $g(500)$ and $g(1,000,000)$ and explain what they represent in this context.

   b. Use the function g to determine the average cost per gallon the airport owner invests to supply the number of gallons of fuel x given in the table below.

x	Cost $f(x)$ (dollars)	Average Cost $g(x)$ (dollars per gallon)
1,000		
10,000		
100,000		
1,000,000		
10,000,000		
100,000,000		

   c. As the number of gallons of fuel supplied increases, how does the average cost per gallon of fuel change?

   d. What is the change in the average cost per gallon as the number of gallons of fuel supplied increases from 1,000,000 to 10,000,000?

   e. What value does the output of g approach as the number of gallons supplied increases without bound? What does this information convey about the co-variation of quantities in the context of the problem?

   f. If the number of gallons of fuel supplied by the jet fuel company continues to increase, will the average cost per gallon reach a minimum value? Explain.

*5. *Recall the following context from Investigation 1 of this module.* Five gallons of liquid flavoring is poured into a large vat. Water will be added to the vat and mixed with the flavoring to produce a drink that will be bottled and sold.

We defined a function f to relate the ratio of flavoring to water $R = f(x)$ in the mixture to the total volume of the mixture x in gallons by $f(x) = \frac{5}{x-5}$.

a. How does the ratio of flavoring to water change as $x \to \infty$ (as x increases without bound)?

b. Is there a minimum value for the ratio of flavoring to water $f(x)$? If so, state the minimum value. If not, explain why there is no minimum value.

c. Use a calculator or graphing software to graph f. With your group or as a class, discuss how the graph supports your answer to (b).

d. Complete the following statement. As $x \to \infty$, $f(x) \to$ _____.

6. A national park research team noticed a dramatic reduction in the deer population in a 150,000-acre protected area. In order to increase the population of deer, the park services introduced 125 additional deer into the area. The researchers' population model predicts that the expected number of deer $f(t)$ is modeled by the function f, defined by $f(t) = \frac{50(2t+16)}{0.0075t+4}$ where t, the time since the 125 deer were introduced, is measured in years.

a. What was the deer population previous to the addition of the 125 deer? Explain how you determined this value.

b. Find the population when i) $t = 7$ years; ii) $t = 18$ years; and iii) $t = 110$ years.

c. When will the population of deer reach 760?

d. What expression can be used to estimate the long-term behavior of the function's values?

e. The research team's model predicts that the total number of deer that can be supported by the 150,000-acre area is limited. What is the maximum number of deer that the research team expected the 150,000-acre area to support? Discuss your approach with your group or as a class.

7. Given the function f defined by $f(x) = \frac{1}{x-8}$, complete the following.

   a. Fill in the tables of values.

x	$x-8$	$f(x) = \frac{1}{x-8}$
$-1{,}000{,}000$		
$-10{,}000$		
-100		

x	$x-8$	$f(x) = \frac{1}{x-8}$
100		
$10{,}000$		
$1{,}000{,}000$		

   b. How does the output of f change as $x \to \infty$? Why does this happen?

   c. How does the output of f change as $x \to -\infty$? Why does this happen?

   d. Using a calculator, graph f. Explain how the graph supports your conclusions above.

*8. Given the function f defined by $f(x) = \frac{3x}{x+4}$, answer the questions below.

   a. Fill in the table of values.

x	$3x$	$x+4$	$f(x) = \frac{3x}{x+4}$
$-1{,}000{,}000$			
$-10{,}000$			
-100			
100			
$10{,}000$			
$1{,}000{,}000$			

   b. As x increases without bound ($x \to \infty$) how does $f(x)$ vary? Explain.

   c. As x decreases without bound ($x \to -\infty$) how does $f(x)$ vary? Explain.

   d. Using a calculator, graph f. Explain how the graph supports your conclusions in parts (b) and (c) above.

Horizontal Asymptotes

A ***horizontal asymptote*** exists if a horizontal line $y = a$ can be used to estimate the end behavior of a function. If $f(x) \to a$ as $x \to \infty$ or $x \to -\infty$ then we say that $y = a$ is a horizontal asymptote for f.

Note that there are other possible end behaviors for functions (including rational functions). Not every rational function will have a horizontal asymptote.

9. Determine any horizontal asymptotes from exercises you have completed thus far. Then discuss with your group or as a class what these asymptotes convey about how x and $f(x)$ change together as $x \to \pm\infty$.

*10. For each given function, do the following.
    i. Write a variable expression that can be used to estimate the function's value as $x \to \pm\infty$. Simplify the expression if possible. Use a form similar to, "As $x \to \pm\infty$, $f(x) \to$ _____."
    ii. Write the function's horizontal asymptote (if one exists).
    iii. Verify your work by graphing the function using a graphing calculator or graphing software.

a. $f(x) = \dfrac{8}{x+1}$ b. $g(x) = \dfrac{x}{2x-8}$ c. $h(x) = \dfrac{6x^2 - 24}{2(x-3)(x+1)}$

d. $m(x) = \dfrac{x+3}{x^2 + 4x - 5}$ c. $n(x) = \dfrac{x^5 + 1}{x^2 + 1}$ f. $c(x) = \dfrac{2(x+6)}{x+6}$

*11. Generalizing the reasoning in this investigation, any polynomial expression
$p(x) = a_n x^n + a_{n-1} x^{n-1} + a_{n-2} x^{n-2} + \ldots + a_1 x + a_0$ can be well-estimated by the variable expression $a_n x^n$
as x increases or decreases without bound. That is, as $x \to \pm\infty$, $p(x) \to a_n x^n$.

Therefore, a rational function $r(x) = \dfrac{p(x)}{q(x)} = \dfrac{a_n x^n + a_{n-1} x^{n-1} + a_{n-2} x^{n-2} + \ldots + a_1 x + a_0}{b_m x^m + b_{m-1} x^{m-1} + b_{m-2} x^{m-2} + \ldots + b_1 x + b_0}$ can be well-
estimated by $\frac{a_n x^n}{b_m x^m}$ as x increases or decreases without bound. That is, as $x \to \pm\infty$, $r(x) \to \frac{a_n x^n}{b_m x^m}$.

a. Under what conditions does a rational function have no horizontal asymptote? Explain and then give an example.

b. Under what conditions does a rational function have a horizontal asymptote at $y = 0$? Explain and then given an example.

c. Under what conditions does a rational function have a horizontal asymptote that is NOT $y = 0$? In this case, how can you determine the horizontal asymptote by looking at the polynomial functions in the numerator and denominator? Explain and give an example.

Use the four rational functions defined in (i-iv) to complete Exercises #1-7.

i) $f(x) = \dfrac{x-3}{x+2}$ *ii) $g(x) = \dfrac{3x^2}{(x-1)(x-3)}$ *iii) $h(x) = \dfrac{x^2+1}{x-2}$ *iv) $k(x) = \dfrac{5x}{x^2-4}$

1. a. Find the real zeros (or roots) of each function.
   i)   ii)   iii)   iv)

   b. Describe your method for determining the zeros and what they represent.

2. a. Determine the vertical intercept of each function and explain what this intercept represents.
   i)   ii)   iii)   iv)

   b. Describe your method for determining the vertical intercepts and what they represent.

3. a. What is the domain of each function?
   i)   ii)   iii)   iv)

   b. For what values of x that are excluded from a function's domain does the graph of the function have a hole instead of a vertical asymptote? Explain how you know that the graph has a hole.

4. a. Using what you learned in Investigation 2, fill in the blanks below.

   i) As $x \to \infty$, $f(x) \to$ _____   ii) As $x \to \infty$, $g(x) \to$ _____
      As $x \to -\infty$, $f(x) \to$ _____   As $x \to -\infty$, $g(x) \to$ _____
      Horizontal Asymptote: _____   Horizontal Asymptote: _____

   iii) As $x \to \infty$, $h(x) \to$ _____   iv) As $x \to \infty$, $k(x) \to$ _____
      As $x \to -\infty$, $h(x) \to$ _____   As $x \to -\infty$, $k(x) \to$ _____
      Horizontal Asymptote: _____   Horizontal Asymptote: _____

b. Use the information from your answers above to sketch a
graph of each of the four functions and then check your work
with a graphing calculator. (*Remember that if you are ever
unclear about how a function behaves on a given interval,
you can evaluate the function for values of x.*).

i)

**If your sketch was incorrect, think carefully about why your
graph was incorrect.**

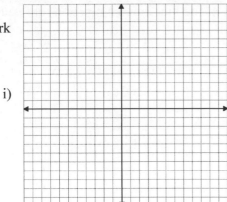

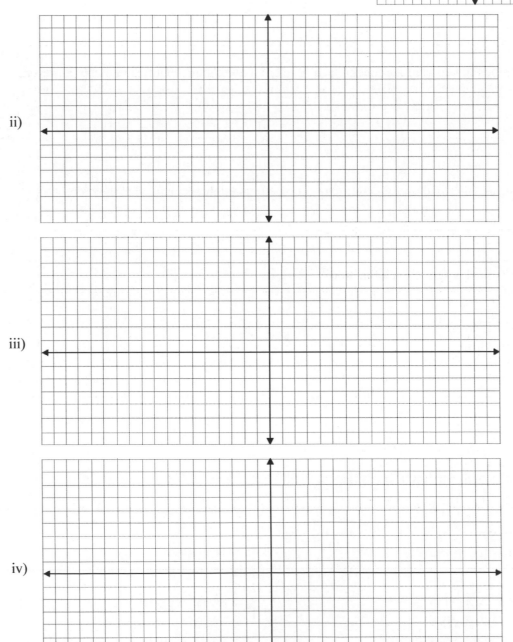

ii)

iii)

iv)

5. Sketch a graph of the following rational functions without using a graphing calculator by determining each function's: i) roots; ii) y-intercept; iii) vertical asymptote(s); iv) horizontal asymptote(s); and (v) the sign of the function on intervals of the function's domain.
   (Remember that if you are ever unclear about the value of a function or how the function behaves on a given interval, you can evaluate the function for values of x.).

   a. $g(x) = \frac{x+7}{x-5}$

   b. $h(x) = \frac{x^2-9}{x+4}$

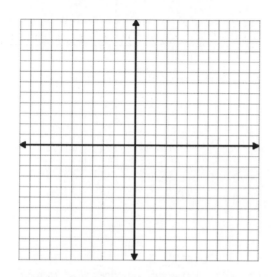
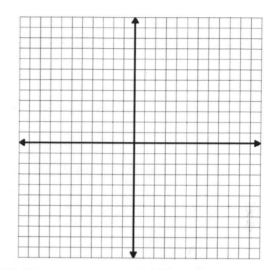

Limit Notation

Limit Notation is a concise way to communicate behavior about a function. If the output values are approaching a specific value as the input value approaches a specific value (or increases or decreases without bound) we say that value is a *limit* of the function values.

For example, consider the function $f(x) = \frac{x-3}{x+2}$. As x increases without bound ($x \to \infty$) the function values approach 1. Therefore, we can write "$\lim\limits_{x \to \infty} f(x) = 1$" (read: "As x increases without bound, the function values approach 1.") In addition, as x decreases without bound the function values also approach 1. We can write this as $\lim\limits_{x \to -\infty} f(x) = 1$.

There is a vertical asymptote at $x = -2$, and by doing a bit of work we can show that $f(x)$ decreases without bound as $x \to -2^-$. Since the output values do not approach a specific value a limit does not exist, or $\lim\limits_{x \to -2^-} f(x)$ DNE. Similarly $\lim\limits_{x \to -2^+} f(x)$ DNE since as $x \to -2^+$, $f(x)$ increases without bound.

---

6. Use the following functions to fill in the blanks below. (Note: If a value does not exist, write DNE)

   $f(x) = \frac{x+3}{x+6}$ $g(x) = \frac{3x^2}{(x-1)(x-3)}$ $h(x) - \frac{x^2+1}{x-2}$ $k(x) = \frac{5x}{x^2-4}$

   a. $\lim\limits_{x \to \infty} f(x) = $ _____

   $\lim\limits_{x \to -\infty} f(x) = $ _____

   $\lim\limits_{x \to -6^-} f(x) = $ _____

   $\lim\limits_{x \to -6^+} f(x) = $ _____

   b. $\lim\limits_{x \to \infty} g(x) = $ _____

   $\lim\limits_{x \to 1^-} g(x) = $ _____

   $\lim\limits_{x \to 1^+} g(x) = $ _____

   $\lim\limits_{x \to 3^+} g(x) = $ _____

   c. $\lim\limits_{x \to \infty} h(x) = $ _____

   $\lim\limits_{x \to -\infty} h(x) = $ _____

   $\lim\limits_{x \to 2^-} h(x) = $ _____

   $\lim\limits_{x \to 2^+} h(x) = $ _____

   d. $\lim\limits_{x \to -\infty} k(x) = $ _____

   $\lim\limits_{x \to -2^-} k(x) = $ _____

   $\lim\limits_{x \to 2^-} k(x) = $ _____

   $\lim\limits_{x \to 2^+} k(x) = $ _____

---

7. For each part you are given information about a rational function. Use the information to sketch a possible graph of the function.

a. $\lim_{x \to \infty} f(x) = -3$

$\lim_{x \to -\infty} f(x) = -3$

$\lim_{x \to 2^-} f(x)$ DNE

As $x \to 2^-$, $f(x)$ increases without bound.

$\lim_{x \to 2^+} f(x)$ DNE

As $x \to 2^+$, $f(x)$ decreases without bound.

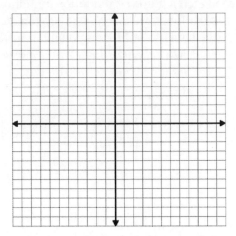

b. $\lim_{x \to \infty} g(x)$ DNE

As $x \to \infty$, $g(x)$ increases without bound.

$\lim_{x \to -\infty} g(x)$ DNE

As $x \to -\infty$, $g(x)$ decreases without bound.

$\lim_{x \to 4^-} g(x)$ DNE

As $x \to 4^-$, $g(x)$ decreases without bound.

$\lim_{x \to 4^+} g(x)$ DNE

As $x \to 4^+$, $g(x)$ increases without bound.

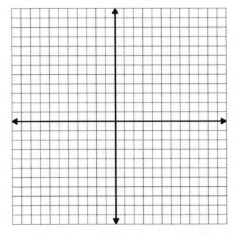

c. $\lim_{x \to \infty} h(x) = 2$ and $\lim_{x \to -\infty} h(x) = 2$

$\lim_{x \to -3^-} h(x)$ DNE

As $x \to -3^-$, $h(x)$ increases without bound.

$\lim_{x \to -3^+} h(x)$ DNE

As $x \to -3^+$, $h(x)$ decreases without bound.

$\lim_{x \to 1^-} f(x)$ DNE

As $x \to 1^-$, $h(x)$ decreases without bound.

$\lim_{x \to 1^+} f(x)$ DNE

As $x \to 1^+$, $h(x)$ increases without bound.

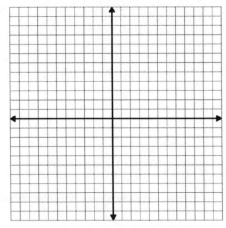

1. Given the graph of the function f determine:
    a.   i. the root(s) of f;   ii. the y-intercept of f;

         iii. the interval(s) on which f is increasing;   iv. the interval(s) on which f is decreasing;

         v. the intervals on which $f(x) > 0$;   vi. the interval(s) on which $f(x) < 0$;

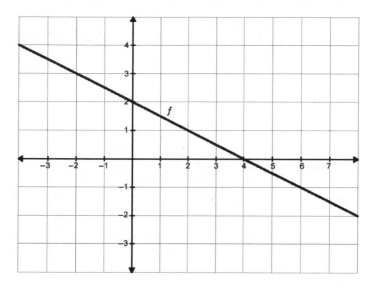

    b.   Consider the function g defined by $g(x) = \frac{1}{f(x)}$ and think about how x and $g(x)$ change together to answer the following.
         i.   What does the value of $g(x)$ approach as $x \to 4^-$?
         ii. What does the value of $g(x)$ approach as $x \to 4^+$?
         iii. What does the value of $g(x)$ approach as $x \to +\infty$ (x increases without bound)?
         iv. What does the value of $g(x)$ approach as $x \to -\infty$ (x decreases without bound)?
         v.   Sketch a graph of g on the axes below.

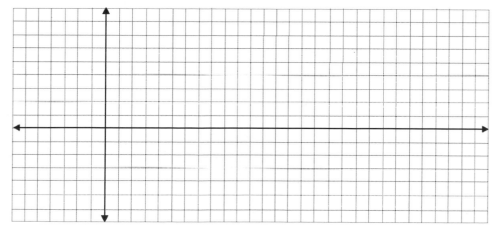

    c.   How would the graph of g change if g were defined by $g(x) = \frac{10}{f(x)}$ instead of $g(x) = \frac{1}{f(x)}$? Explain your reasoning.

d.   How would the graph of g change if g were defined by $g(x) = \frac{1}{f(x)+1}$ instead of $g(x) = \frac{1}{f(x)}$? Explain your reasoning.

4.   Given the graph of f sketch the graph of g defined by $g(x) = \frac{5}{f(x)}$.

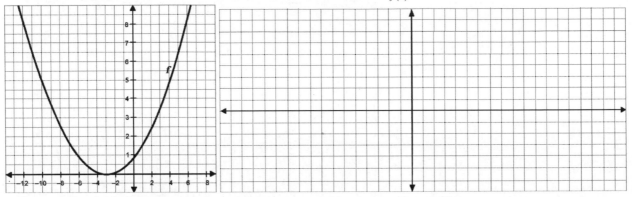

5. Given the graphs of f and g sketch the graph of h defined by $h(x) = \frac{f(x)}{g(x)}$.

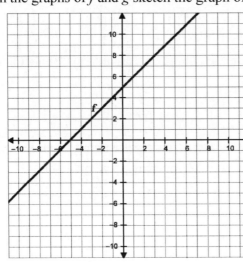
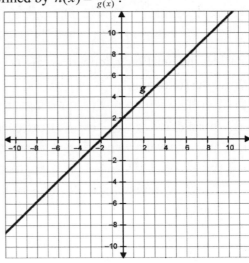

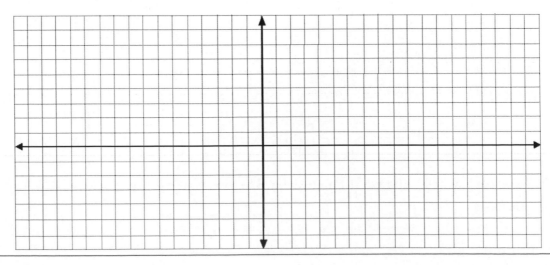

0. OPTIONAL CONTENT: SYMBOLIC OPERATIONS

In Exercises #1-6, rewrite each function in simplified form and then describe the function's domain.

1. $f(x) = \frac{x^2 - 5x + 6}{3x^2 - 15x + 18}$

2. $f(x) = \frac{3x^4 - 9x^2}{9x^3 + 27x^2}$

3. $f(x) = \frac{x^3 + x^2 - 2x}{-2 + x^2 + x}$

4. $f(x) = \frac{4x^2 - 20x - 56}{2x^2 - 22x + 56}$

5. $f(x) = \frac{2x^2 + 7x - 15}{2x^2 - x - 3}$

6. $f(x) = \frac{-x^2 + 6x - 8}{x^2 - x - 12}$

In Exercises #7-12, perform the specified arithmetic operations and simplify the result if possible.

7. $\frac{4x}{x+4} + \frac{x-6}{x-9}$

8. $\frac{5x^2}{x^3 + 1} + \frac{1}{x}$

9. $\frac{x^2 - 1}{x - 8} - \frac{x + 8}{7}$

10. $\frac{x^2}{(x+8)^4} \cdot \frac{(x+8)^6}{x^3(2x-1)}$

11. $\frac{2x^2 - x - 1}{x+3} \cdot \frac{x+3}{x-1}$

12. $\frac{(4x-2)^3}{(x+3)^2} \div \frac{(4x-2)^5}{(x-3)^2}$

In Exercises #13-18, rewrite the sums and differences as products by factoring the polynomials. For example, given $4(x+6) + (x^2 - 2)(x+6)$, both terms have $x+6$ as a common factor, so we can rewrite it as $(x+6)\big((4) + (x^2 - 2)\big)$ by factoring out $x+6$, or $(x+6)(4 + x^2 - 2)$ or $(x+6)(x^2 + 2)$ simplified.

13. $10(x+14) - 2x(x+14)$

14. $x^2(x^2 - 1) + 13(x^2 - 1)$

15. $3x(x^2 + 5x - 14) - (x^2 + 5x - 14)$

16. $(2x^2 + 6)^2(x^2 - 2x)^3 + (2x^2 + 6)^3(x^2 - 2x)^2$

17. $(x+1)^2(x+2)^3(x+3)^4 + (x+1)^3(x+2)^4(x+3)^5$

18. $(2x)^{1/2} - (2x)^{5/2}$

I. INTRODUCTION TO RATIONAL FUNCTIONS AND VERTICAL ASYMPTOTES (Text: S1, 2, 4, 6)

19. The given table shows values of two functions. Function f inputs the year and outputs the average annual compensation (in dollars) for corporate CEOs in the U.S. Function g inputs the year and outputs the average annual compensation (in dollars) for non-manager corporate employees. *Note that the values are adjusted for inflation. Data collected from the Economy Policy Institute.*

year, t	average CEO compensation (dollars) $f(t)$	average non-manager compensation (dollars) $g(t)$
1965	819,000	39,500
1973	1,069,000	46,400
1978	1,463,000	47,200
1989	2,724,000	44,700
1995	5,768,000	45,600
2000	20,172,000	47,900
2010	12,466,000	52,700
2013	15,175,000	52,100

    a. Evaluate $f(2000)$ and $g(1989)$ and explain what they represent.

    b. The output of function h is the value of the ratio $\frac{f(t)}{g(t)}$. That is, $h(t) = \frac{f(t)}{g(t)}$. Evaluate $h(1978)$ and $h(1995)$ and explain what the values represent in this context.

    c. As t increases from 1978 to 1995, did the values of $f(t)$ and $g(t)$ increase, decrease, or remain constant? What does that tell us about this context?

    d. As t increases from 1978 to 1995, did the value of $h(t)$ increase, decrease, or remain constant? What does that tell us about this context?

    e. What would it mean for $h(t) = 10$ for some value of t? What would it mean for $h(t) = 1$ for some value of t?

    f. Is it possible for $h(t)$ to be less than 1 for some value of t?

e.   Use the information in the table to create a table of values for function *h*. Describe the behavior of *h* as *t* increases and what that tells you about this context.

20. A weight loss center is so confident in their program that they created a special pricing plan. Joining the program costs a one-time fee of $49.99, and members don't have to pay anything more unless they lose weight. However, members pay the center $2.99 per pound they lose while in the program.

a.   Define a function *f* to express the cost *C* for someone joining the program and losing *x* pounds. What is the practical domain of this function?

b.   Suppose we are interested in determining the average cost per pound lost for someone in the program. (*Note that average cost is NOT an average rate of change*.) Describe how to determine the average cost per pound lost when *x* pounds are lost by a member in the program. (*What calculation determines the average cost and why?*)

c.   Complete the given table and then determine the average cost per pound lost for a member losing *x* pounds while in the program.

d.   Define a function *g* that relates the member's average cost per pound lost (measured in dollars) as a function of the number of pounds lost.

e.   When *x* is positive and getting very close to 0 (which we write as $x \to 0^+$), what happens to the average cost per pound? Why?

f.   Does the average cost per pound have a maximum value? Explain.

x	Cost $f(x)$	Average Cost $g(x)$
0.01		
0.1		
1		
10		
100		

21. A beverage company has just completed brewing a large batch of tea (1500 gallons). They will add sweetening to the tea, then bottle it and prepare it for distribution.

a.   Suppose 10 gallons of corn syrup is added to the mixture as a sweetener. What is the ratio of tea to corn syrup in the mixture? What if 50 gallons of corn syrup is added instead? 200 gallons?

b.   Define a function *f* whose input *x* is the amount of corn syrup added (in gallons) and whose output *R* is the ratio of tea to corn syrup in the mixture.

c.   What is the practical domain for the function you defined?

d.   Complete the given table of values for this function.

e.   What happens to *R* as $x \to 0^+$? Why does this make sense?

x	$R = f(x)$
0.001	
0.01	
0.1	
1	

22. A cylindrical container is designed to carry liquid that must be insulated (*see diagram*). The interior cylinder will hold the liquid, and the insulation must be 2 inches thick.

a.   If the radius of the entire container (interior cylinder and insulation) is 4.5 inches, what is the radius of the interior cylinder? What if the radius of the entire container is 5.9 inches? *x* inches?

b.   The formula for the volume of a cylinder is given by $V = \pi r^2 h$. Suppose the container must hold exactly 200 in³ of liquid. How tall must the container be if the radius of the entire container is 4.5 inches and the interior cylinder holds exactly 200 in³? What if the radius of the entire container is 5.9 inches?

c.   Define a function *f* whose input *x* is the radius of the entire package (in inches) and whose output *h* is the height of the package (in inches) necessary so that the interior container holds 200 in³ of liquid.

d.   What happens to the value of *h* as $x \to 2^+$? What does this mean in the context of this problem?

23. Given that $f(x) = \frac{x}{x-9}$, answer the questions below.

a.   Complete the given tables of values. Show the calculations that provided your answers.

b.   How do the output values of $f(x)$ change as $x \to 9^-$? Why?

x	$f(x)$
8	
8.9	
8.99	
8.999	

x	$f(x)$
9.001	
9.01	
9.1	
10	

   c. How do the output values of $f(x)$ change as $x \to 9^+$? Why?

   d. Using a calculator, graph f. Explain how the graph supports your conclusions above.

   e. What changes if the definition of f becomes $f(x) = -\frac{x}{x-9}$? Why does this happen?

   f. What changes if the definition of f becomes $f(x) = \frac{x}{x+9}$? Why does this happen?

24. Given that $f(x) = \frac{2}{x+7}$, complete the following.

   a. How do the output values of f change as $x \to -7^-$?

   b. How do the output values of f change as $x \to -7^+$?

   c. Using a calculator, graph f. Explain how the graph supports your conclusions in parts (a) and (b).

   d. What changes if the definition of f becomes $f(x) = -\frac{2}{x+7}$? Why does this happen?

25. Given that $g(x) = \frac{5x}{(x+1)(x-3)}$, complete the following.

   a. How do the output values of g change as $x \to -1^-$? As $x \to -1^+$?

   b. How do the output values of g change as $x \to 3^-$? As $x \to 3^+$?

   c. Using a calculator, graph f. Explain how the graph supports your conclusions in parts (a) and (b).

26. For each of the following functions, predict where a vertical asymptote will exist and then test your prediction using a graphing calculator.

   a. $f(x) = \frac{x-10}{x+10}$ b. $g(x) = \frac{x^2}{5x-9}$ c. $h(x) = \frac{x}{x^2+2}$ d. $m(x) = \frac{6}{x^2-1}$ e. $n(x) = \frac{5x+5}{x+1}$ f. $p(x) = \frac{6-x}{(x-4)(x-5)}$

II. END BEHAVIOR OF RATIONAL FUNCTIONS (Text: S1, 2, 3, 5)

27. A beverage company has just completed brewing a large batch of tea (1500 gallons). They will add corn syrup to the tea, then bottle it and prepare it for distribution. The function f relates the ratio R of tea to corn syrup in the beverage when x gallons of corn syrup are added and is defined by $f(x) = \frac{1500}{x}$ with $R = f(x)$.

x	$R = f(x)$
100	
10,000	
1,000,000	
10,000,000	

   a. Complete the given table of values for this function.

   b. What happens to R as x increases without bound?

   c. Is it practical in this context to allow x to increase without bound? Explain.

28. A cylindrical container is being designed to carry liquid that must be insulated (see diagram below). The interior cylinder will hold the liquid, and the insulation must be 2 inches thick. Function f is defined by $f(x) = \frac{200}{\pi(x-2)^2}$ with $h = f(x)$ where x

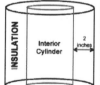

x	$h = f(x)$
10	
100	
1,000	
10,000	

is the radius of the entire package (in inches) and h is the height of the package (in inches) necessary so that the interior container holds 200 in³ of liquid.

   a. Complete the given table of values.

   b. What happens to the value of h as x increases without bound? What does this mean in the context of this problem?

29. Clark's Soda Company incurred a start-up cost of $1276 for equipment to produce a new soda flavor. The cost of producing the drink is $0.26 per can. The company sells the soda for $0.75 per can. (*Assume that Clark's Soda Company is able to sell every can produced.*)

   a. Define a function f to determine the cost (including the start-up cost) measured in dollars to produce x cans of soda.

   b. Define a function g to determine the revenue (measured in dollars) generated from selling x cans of soda.

   c. Define a function h to determine the profit (measured in dollars) from selling x cans of soda (recall that profit = revenue − cost).

   d. Define a function A to determine the average cost per can of producing the soda.

   e. How does $A(x)$ change as the number of cans produced gets larger and larger? Will the average cost $A(x)$ ever reach a minimum value? Explain.

30. A salt cell is cleaned using a mixture of water and acid. There are currently 10 liters of water in a bucket. A technician adds varying amounts of acid to the 10 liters of water.

- Let x = the number of liters of acid added to the 10 liters of water
- Let A = the level of acidity of the mixture (or the percentage of the solution that is acid)

   a. Complete the given table of the mixture's acidity (*measured as a percentage of the total*) as acid is added to the 10 liters of water.

   b. Define a function to determine the acidity of the mixture (measured as a percentage) in terms of the number of liters of acid x that has been added.

   c. Use a calculator to graph f.

   d. Describe how the acidity of the mixture changes as the number of liters of acid added to the solution increases without bound.

   e. Is it possible for the mixture to ever reach 100% acid? Explain.

   f. If the water is too acidic, it can damage the salt cell. If the water is not acidic enough, the mixture will not be practical in cleaning the cell. It turns out the mixture is most efficient if the acidity is between 40% and 60%. What is the range of liters of acid that should be added to the bucket of water so that the mixture is most efficient?

   g. Define a function that accepts the desired percentage of acid as input and determines the number of liters of acid that should be added to the water as output.

liters of acid x	acitidy of mixture, A (as a percent)
0	
0.5	
1.0	
1.5	
2.0	
2.5	
3.0	
3.5	

31. A large mosquito repellent manufacturer produces a repellent using DEET (the most common active ingredient in insect repellents) and a moisturizer lotion. A tank for mixing the repellent is filled with a steady flow of the lotion and DEET. Soon after the filling began, the mixture was sampled and it was determined that there were 97 ounces of lotion and 4 ounces of DEET in the tank (let the time of the first sample be $t = 0$ where time is measured in a number of seconds). Additional readings revealed that the lotion was flowing in at a constant rate of 12 ounces per second and the DEET was flowing in at a constant rate of 6.4 ounces per second. The company needs to monitor the percentage of DEET in their lotion.

   a. Define a function f to determine the number of ounces of DEET in the tank as a function of time (measured in seconds).

   b. Define a function g to determine the *total* number of ounces of the mixture (both lotion and DEET) that are in the tank as a function of time (measured in seconds).

   c. Define a function h to determine the percentage of DEET in the mixture as a function of time (measured in seconds).

   d. Complete the given table. Round your answers to the nearest ten-thousandths.

   e. Describe how the percentage of DEET in the solution changes as time increases. Will the percentage of DEET in the mixture ever reach a maximum? Explain.

   f. Describe the end-behavior (as t increases without bound) of h.

   g. Use a calculator to graph h, then explain how the graph supports your answer in part (f).

Number of seconds elapsed, t	Percentage of DEET in the mixture, $h(t)$
0	
50	
100	
200	
500	

32. For each given function, state the horizontal asymptote (if one exists).

a. $f(x) = \frac{x}{2x-3}$ b. $g(x) = \frac{x^2}{x+5}$ c. $h(x) = \frac{4x+1}{2x-10}$ d. $m(x) = \frac{x^2-2}{x^2+3x+2}$ e. $n(x) = \frac{17x+200}{10x^3-100x^2}$

f. $p(x) = \frac{5x^2}{x(x-4)}$ g. $q(x) = \frac{2x^3+7}{5x^4-10}$ h. $r(x) = \frac{4x^2+x+11}{(3x+1)(2x-3)}$ i. $w(x) = \frac{(x+9)(x-5)}{(x+6)(x+2)(x-3)}$

III. GRAPHING RATIONAL FUNCTIONS AND UNDERSTANDING LIMITS (Text: S3, 4, 5, 6)

33. Which of the following best describes the behavior of the function f defined by $f(x) = \frac{1}{(x-2)^2}$?
    Provide a rationale for your answer.
    a. As the value of x approaches positive infinity, the value of f decreases without bound.
    b. As the value of x approaches positive infinity, the value of f increases without bound.
    c. As the value of x approaches positive infinity, the value of f approaches 0.
    d. As the value of x approaches 2, the value of f approaches 0.
    e. (a) and (c)

In Exercises #34-42, identify the x-intercepts (roots), y-intercept, horizontal asymptotes, vertical asymptotes, and the function's domain. (*State DNE in cases when an intercept or asymptote does not exist.*) Use this information to sketch a graph of the function.

34. $a(x) = \frac{3}{x-7}$ 35. $b(x) = \frac{x}{2x+6}$ 36. $d(x) = \frac{9x}{3x+3}$ 37. $f(x) = \frac{-x+3}{2x-1}$ 38. $g(x) = \frac{9x^2-144}{x^2-1}$

39. $h(x) = \frac{14x}{3x-4}$ 40. $k(x) = \frac{x-11}{x^2-5x+6}$ 41. $p(x) = \frac{x(x+2)}{4x+1}$ 42. $q(x) = \frac{x^2+2x-3}{x^2-1}$

For Exercises #43-44, use the following graph of f.

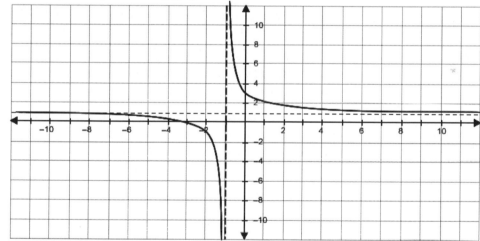

43. Use the graph of f to determine the following limits. If the limit does not exist, write DNE.
    a. $\lim\limits_{x \to \infty} f(x)$ b. $\lim\limits_{x \to -\infty} f(x)$ c. $\lim\limits_{x \to 2^+} f(x)$ d. $\lim\limits_{x \to 2^-} f(x)$

    e. $\lim\limits_{x \to 2} f(x)$ f. $\lim\limits_{x \to -1^+} f(x)$ g. $\lim\limits_{x \to -1^-} f(x)$ h. $\lim\limits_{x \to -1} f(x)$

44. Using the graph of f, identify the vertical and horizontal intercepts, the vertical and horizontal asymptotes, and the domain and range of the function. Then, determine a possible rule for the function f.

For Exercises #45-46, use the following graph of *g*.

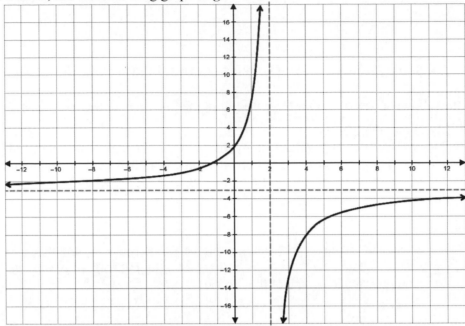

45. Use the graph of *g* to determine the following limits. If the limit does not exist, write DNE.

   a. $\displaystyle\lim_{x\to\infty} g(x)$
   b. $\displaystyle\lim_{x\to-\infty} g(x)$
   c. $\displaystyle\lim_{x\to2^+} g(x)$
   d. $\displaystyle\lim_{x\to2^-} g(x)$

   e. $\displaystyle\lim_{x\to2} g(x)$
   f. $\displaystyle\lim_{x\to6^+} g(x)$
   g. $\displaystyle\lim_{x\to6^-} g(x)$
   h. $\displaystyle\lim_{x\to6} g(x)$

46. Use the graph of *g* to identify the vertical and horizontal intercepts, the vertical and horizontal asymptotes, and the domain and range of the function. Then, determine a possible rule for the function *g* (i.e., represent the function algebraically).

For Exercises #47-48, use the following graph of *h*.

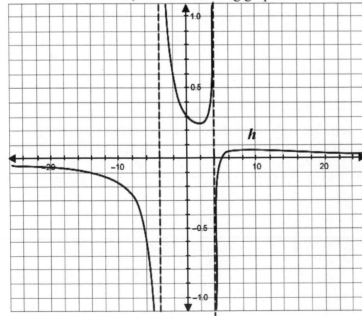

47. Use the graph of *h* to determine the following limits. If the limit does not exist, write DNE.

   a. $\displaystyle\lim_{x\to\infty} h(x)$
   b. $\displaystyle\lim_{x\to-\infty} h(x)$

   c. $\displaystyle\lim_{x\to-4^+} h(x)$
   d. $\displaystyle\lim_{x\to-4^-} h(x)$

   e. $\displaystyle\lim_{x\to-4} h(x)$
   f. $\displaystyle\lim_{x\to4^+} h(x)$

   g. $\displaystyle\lim_{x\to4^-} h(x)$
   h. $\displaystyle\lim_{x\to4} h(x)$

48. Use the graph of *h* to identify the vertical and horizontal intercepts, the vertical and horizontal asymptotes, and the domain and range of the function. Then, determine a possible rule for the function *h*.

For Exercises 49 and 50 using the following table of values for the rational functions f, g, and h.

x	-10000	-1000	-100	100	1000	10000
$f(x)$	-10001.9997	-1001.9970	-101.9694	97.9706	997.9970	9997.9997
$g(x)$	-6.0004	-6.0040	-6.0404	-5.9604	-5.9960	-5.9996
$h(x)$	-0.0005	-0.0050	-0.0506	0.0496	0.0050	0.0005

49. Based on the table of values above, explain the end behavior of each function.

    a. $\lim\limits_{x \to \infty} f(x)$ b. $\lim\limits_{x \to -\infty} f(x)$ c. $\lim\limits_{x \to \infty} g(x)$ d. $\lim\limits_{x \to -\infty} g(x)$

    e. $\lim\limits_{x \to \infty} h(x)$ f. $\lim\limits_{x \to \infty} h(x)$

50. Consider the rule that defines the functions f, g, and h. How does the degree of the numerator compare to the degree of the denominator (e.g., greater than, less than, or equal to)?

51. Use the given information about function f to complete parts (a) through (d).

    • $\lim\limits_{x \to \infty} f(x) = 0$ and $\lim\limits_{x \to -\infty} f(x) = 0$

    • $\lim\limits_{x \to 6^+} f(x)$ DNE because $f(x)$ increases without bound as $x \to 6^+$

    • $\lim\limits_{x \to 6^+} f(x)$ DNE because $f(x)$ decreases without bound as $x \to 6^-$

    •   The y-intercept is $\frac{5}{36}$ (so $\left(0, \frac{5}{36}\right)$ is a point on the graph).

    •   The x-intercept is 5 (so $(5, 0)$ is a point on the graph).

    a. Identify the vertical and horizontal asymptotes for f. State DNE if none exist.
    b. What is the domain of f
    c. Find a possible formula for f.
    d. Sketch a graph for f.

52. Use the given information about function g to complete parts (a) through (d).

    • $\lim\limits_{x \to \infty} g(x) = 7$ and $\lim\limits_{x \to -\infty} g(x) = 7$

    • $\lim\limits_{x \to 2^+} g(x)$ DNE because $g(x)$ decreases without bound as $x \to 2^+$

    • $\lim\limits_{x \to 2^-} g(x)$ DNE because $g(x)$ increases without bound as $x \to 2^-$

    •   The y-intercept is $\frac{7}{2}$ (so $\left(0, \frac{7}{2}\right)$ is a point on the graph).

    •   The x intercept is -1 (so $(-1, 0)$ is a point on the graph).

    a. Identify the vertical and horizontal asymptotes for g. State DNE if none exist.
    b. What is the domain of g?
    c. Find a possible formula for g.
    d. Sketch a graph for g.

53. Use the given information about function *h* to complete parts (a) through (d).

- $\lim\limits_{x \to \infty} h(x)$ DNE because $h(x)$ increases without bound as $x \to \infty$

- $\lim\limits_{x \to -\infty} h(x)$ DNE because $h(x)$ increases without bound as $x \to -\infty$

- $\lim\limits_{x \to -(9/2)^+} h(x)$ DNE because $h(x)$ increases without bound as $x \to -(9/2)^+$

- $\lim\limits_{x \to -(9/2)^-} h(x)$ DNE because $h(x)$ increases without bound as $x \to -(9/2)^-$

- The *y*-intercept is $-\frac{10}{9}$ (so $\left(0, -\frac{10}{9}\right)$ is a point on the graph).

- The *x*-intercepts are $-\sqrt{5}$ and $\sqrt{5}$ (so $\left(-\sqrt{5}, 0\right)$ and $\left(\sqrt{5}, 0\right)$ are points on the graph).

  a. Identify the vertical and horizontal asymptotes for *h*. State DNE if none exist.
  b. What is the domain of *h*?
  c. Find a possible formula for *h*.
  d. Create a graph for *h*.

IV. CO-VARIATION OF NUMERATORS AND DENOMINATORS IN RATIONAL FUNCTIONS

54. Use the graph of *f* to answer the questions that follow.

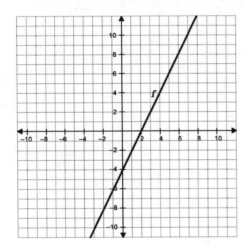

  a. Describe the important characteristics of *f*.
  b. Define a new function *g* as $g(x) = \frac{1}{f(x)}$.
     i. What happens to $g(x)$ as $x \to 2^-$? As $x \to 2^+$?
     ii. What happens to $g(x)$ as *x* increases without bound? As *x* decreases without bound?
     iii. Sketch a graph of *g*.
  c. What, if anything, changes if the definition of *g* changes to become $g(x) = \frac{10}{f(x)}$? Why does this happen?

55. Use the graph of *f* below to answer the questions that follow.

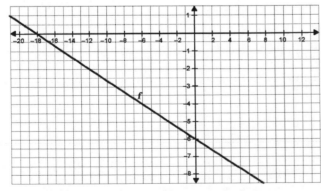

  a. Describe the important characteristics of *f*.
  b. Define a new function *g* as $g(x) = \frac{2}{f(x)}$.
     i. What happens to $g(x)$ as $x \to -18^-$? As $x \to -18^+$?
     ii. What happens to $g(x)$ as *x* increases without bound? As *x* decreases without bound?
     iii. Sketch a graph of *g*.
  c. What, if anything, changes if the definition of *g* changes to become $g(x) = -\frac{2}{f(x)}$? Why?

56. Given the graph of *f* below, sketch the graph of *g* if $g(x) = -\frac{8}{f(x)}$.

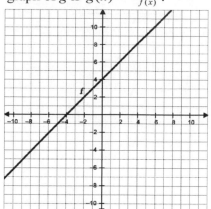

57. Given the graph of *f* below, sketch the graph of *g* if $g(x) = \frac{1}{f(x)}$.

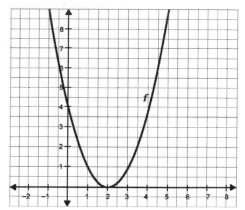

58. Given the graph of *f* below, sketch the graph of *g* if $g(x) = \frac{4}{f(x)}$.

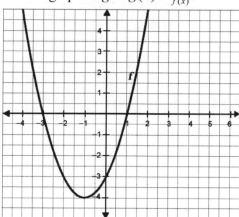

59. Given the graph of *f* below, sketch the graph of *g* if $g(x) = \frac{x}{f(x)}$.

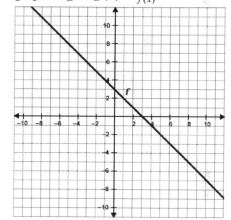

60. Given the graphs of *f* and *g* below, sketch the graph of *h* if $h(x) = \frac{g(x)}{f(x)}$.

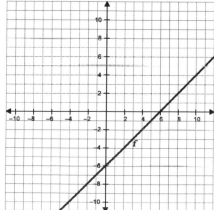
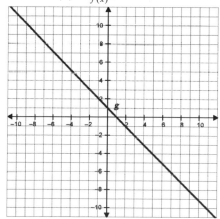

61. Given the graphs of f and g below, sketch the graph of h if $h(x) = \frac{g(x)}{f(x)}$.

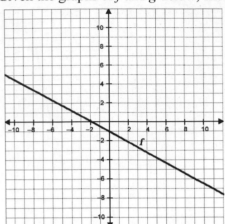
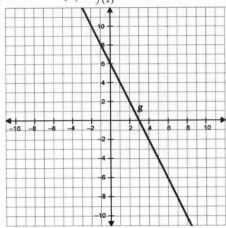

62. Given the graphs of f and g below, sketch the graph of h if $h(x) = \frac{f(x)}{g(x)}$.

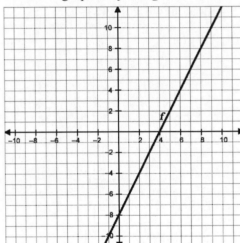
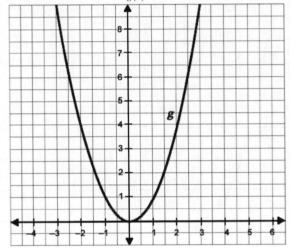

This investigation contains review and practice with important skills and procedures you may need in this module and future modules. Your instructor may assign this investigation as an introduction to the module or may ask you to complete select exercises "just in time" to help you when needed. Alternatively, you can complete these exercises on your own to help review important skills.

Proportional Reasoning
Use this section prior to the module or with/after Investigation 1.

1. When Ann takes her dog on a walk, her dog always takes six steps for each step that Ann takes.
   a. If Ann's dog takes 1,200 steps, how many steps did Ann take?

   b. How will their total number of steps compare when they walk once around the block? Twice around the block? Ten times around the block?

   c. How does the distance Ann walks in one step compare to the distance her dog walks in one step?

2. The perimeter of a square is always four times as long as the length of one of its sides.
   a. Each of the following represents the perimeter for some square (stretched out into a single segment length). For each, draw a segment whose length must be the length of one of the square's sides.

      perimeter of a square
      _____

      perimeter of a square
      _____

      perimeter of a square
      _____

   b. In your own words, explain the important idea demonstrated in part (a).

Use the following information to complete Exercises #3-8.

Lengths
1 cm = 0.39 in
1 ft = 12 in
1 mile = 5,280 ft
1 ft = 0.305 meters

Volume
1 liter = 33.81 fluid ounces
1 liter = 1.05 quarts
1 gallon = 4 quarts
1 quart = 2 pints

3. If I measure the length of a table to be 54 inches, how long is that table in feet? In cm?

4. If I measure the length of a room to be 14 feet, how long is that room in inches? In meters?

5. If I measure John's height to be 2.1 meters, how tall is he in feet? In inches? In miles?

6. A soda bottle contains 2 liters of liquid. What is its volume in fluid ounces? Quarts? Gallons?

7. A glass contains 1 pint of water. What is the volume in gallons? Liters? Fluid ounces?

8. A speed of 56 miles per hour corresponds to a speed of about 90 km per hour. What does that tell us about the relative size of 1 mile compared to 1 km?

9. Jack uses the length of his hand to measure the size of his desk. He finds the top of his desk to be about "4.5 hands by 6.25 hands" in size. Cory uses his hand and finds the same desk to measure about "5.5 hands by 7.5 hands" in size.
   a. In general, what can we say about Cory's hands compared to Jack's hands?

   b. Can we determine approximately the length of Jack's hand is using Cory's hand as a "1 unit ruler"? Explain.

   c. Can we determine approximately the length of Cory's hand is using Jack's hand as a "1 unit ruler"? Explain.

Breaking Up a Circle's Circumference
Use this section prior to the module or with/after Investigations 1 and 2.

The given diagram shows a circle with its circumference broken into five equal arc lengths.

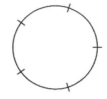

10. For each given circle, make tick marks to break up its circumference into the indicated number of equal arc lengths. *It's okay to approximate the location of the marks but try to be reasonably accurate.*
    a. 8 equal arc lengths

    b. 10 equal arc lengths

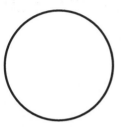

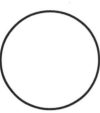

    c. 16 equal arc lengths

    d. 3 equal arc lengths

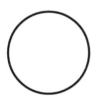

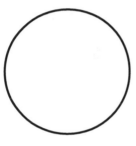

11. Do the following for each circle and set of arcs in Exercise #10.
    i. If circle's circumference is 24.6 inches, how long is each arc?
    ii. What percentage of the circle's circumference is each arc?
    iii. How many of the arcs you created make up half of the circle's circumference?
    a.   b.

    c.   d.

You should recall that a circle's circumference is 2π times as long as its radius. *Since 2π is about 6.28, there are a little more than 6 radii that make up every circle's circumference.*

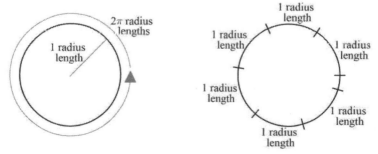

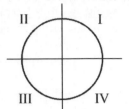

12. Let's place a circle on a coordinate plane (with the circle's center at the origin) and think about starting at the 3 o'clock position (where the circle intersects the horizontal axis to the right). Imagine moving counterclockwise and marking off arcs that are each 1 radius length as you go.
    a. In which quadrant do we stop if we mark off 2 of these arc lengths?

    b. In which quadrant do we stop if we mark off 5 of these arc lengths?

    c. In which quadrant do we stop if we mark off 3.5 of these arc lengths?

    d. In which quadrant do we stop if we mark off 1.1 of these arc lengths?

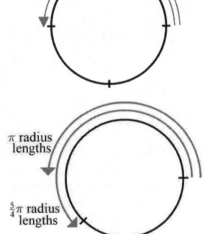

If we break up the circle's circumference into four equal arc lengths, how long is each arc *in radius lengths*?

We know that the circumference is 2π times as long as the radius, so each arc is:

$$\tfrac{1}{4}(2\pi) \text{ radius lengths}$$

$$\tfrac{2}{4}\pi \text{ radius lengths}$$

$$\tfrac{1}{2}\pi \text{ radius lengths}$$

Note that it's common to see $\tfrac{1}{2}\pi$ radius lengths written as $\tfrac{\pi}{2}$. For many students, using the form $\tfrac{1}{2}\pi$ makes it a bit easier to understand the meaning.

For another example, suppose an arc measures $\tfrac{5\pi}{4}$ radius lengths. Can we estimate its length on a circle? First, let's rewrite this as $\tfrac{5}{4}\pi$. In this form it's easy to see that the arc must be $\tfrac{5}{4}$, or 1.25, times as long as π radius lengths. Let's draw this.

13. For each arc length, do the following.
    i.   Rewrite in the form $\tfrac{a}{b}\pi$ where $\tfrac{a}{b}$ is some rational number. For example, rewrite $\tfrac{7\pi}{3}$ as $\tfrac{7}{3}\pi$ or $\tfrac{\pi}{9}$ as $\tfrac{1}{9}\pi$.
    ii. Use the given circle to draw an arc that has the given length. *Start in the 3 o'clock position and move counterclockwise.*
    iii. In what quadrant does the arc's endpoint fall?

    a. $\tfrac{\pi}{3}$ 　　　　　　　　　　b. $\tfrac{7\pi}{10}$

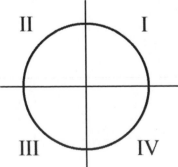 　　　　　　　　　　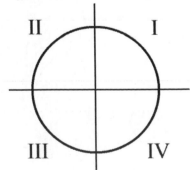

c. $\frac{6\pi}{5}$

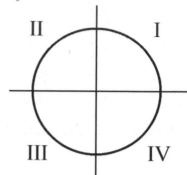

d. $\frac{11\pi}{6}$

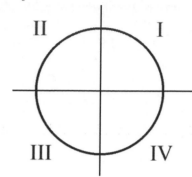

14. For each given circle, make tick marks to break up its circumference into arcs with the indicated length. *It's okay to approximate the location of the marks but try to be reasonably accurate.*

   a. $\frac{\pi}{2}$ radius lengths

   b. $\frac{\pi}{3}$ radius lengths

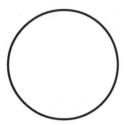

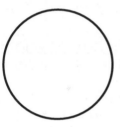

   c. $\frac{\pi}{4}$ radius lengths

   d. $\frac{\pi}{10}$ radius lengths

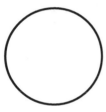

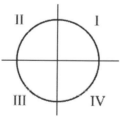

15. Let's place a circle on a coordinate plane (with the circle's center at the origin) and think about starting at the 3 o'clock position and moving counterclockwise.

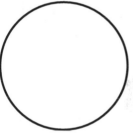

   a. In which quadrant is the endpoint of an arc that measures $\frac{2\pi}{3}$ radius lengths?

   b. In which quadrant is the endpoint of an arc that measures $\frac{11\pi}{10}$ radius lengths?

   c. In which quadrant is the endpoint of an arc that measures $\frac{9\pi}{5}$ radius lengths?

   d. In which quadrant is the endpoint of an arc that measures $\frac{11\pi}{10}$ radius lengths marked off ***clockwise*** from the 3 o'clock position?

Simplifying Expressions and Solving Equations
Use this section prior to the module or with/after Investigation 4.

In Exercises #16-19 decide if each statement is valid. Justify your answer.

16. $4y = 4x + 5$ implies that $y = x + 5$

17. $\frac{ab+c}{a} = b + c$

18. $-5a(-7a + 5) = 35a^2 - 25a$

19. $\frac{2y-6x}{3x} = 2y - 2$

Constant Rate of Change for a Linear Function
Use this section with/after Investigation 8.

In Exercises #20-27 you are given two solutions for a linear function. Determine the function's constant rate of change (slope).

20. $(0, 0)$ and $(1.5, 4.5)$

21. $(0, 0)$ and $(4.2, -16.8)$

22. $(0, 0)$ and $(15, 5)$

23. $(2, 14)$ and $(39.5, 29)$

24. $(8.7, 1)$ and $(28.7, -3)$

25. $(-1.6, -4.7)$ and $(11.9, 13.3)$

26. $(-7.1, 14)$ and $(8.9, -42)$

27. $(-12.2, 24.3)$ and $(-5.18, 16.5)$

*1. a.   What is an angle?

   b.   Which of the given angles has the smallest measure? The largest measure? What quantity did you
        focus on to determine your answers?

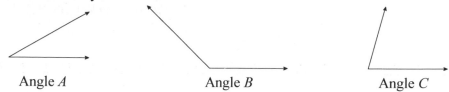

Angle A Angle B Angle C

   c.   Imagine centering circles of equal radii at the vertices of angles A, B, and C. Each of these angles
        subtend, or cut off, a portion of the circle's circumference. Repeat part (b) based on the new
        diagrams.

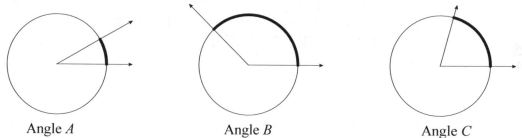

Angle A Angle B Angle C

   d.   Use your response to Part (c) to describe what it means for two angles to have the same measure.

*2. a.   Use a piece of string to measure each angle in units of $1/12^{th}$ of the circumference of the circle
        centered at its vertex. Why might it be useful to measure these angles in units of $1/12^{th}$ of the
        circumference of the circle centered at their respective vertices?

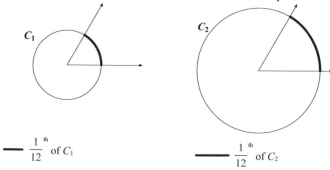

$\underline{\quad\quad}$ $\frac{1}{12}^{th}$ of C_1 $\underline{\quad\quad}$ $\frac{1}{12}^{th}$ of C_2

   b.   Could you measure the two angles from Part (a) by measuring the length of the arcs these angles
        subtend in a standard linear unit like inches or centimeters? Explain.

3.   a.   Approximate the measure of the given angle in unit *m*.

b.   Sketch unit *m* along the subtended arcs in the given figure so that the measure of the angle in unit *m* is maintained. What do you notice about the relationship between the length of unit *m* and the circumference of the circle on which it falls?

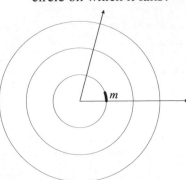

*4.   Make a conjecture about the criterion that a unit of measure for the length of a subtended arc must satisfy and explain why this criterion must be satisfied.

*5.   Reflect on your responses to Exercises #1-4 and then describe what it means to measure an angle. (*Be specific about the quantity one measures when measuring an angle and the type of units one uses to measure this quantity.*)

*6.   Visually estimate the measure of the angles below in the units specified.

    a.   b.   c.

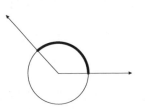

    i.   1/4th of the circumference

i. $1/4^{\text{th}}$ of the circumference   i. $1/4^{\text{th}}$ of the circumference   i. $1/8^{\text{th}}$ of the circumference

ii. $1/6^{\text{th}}$ of the circumference   ii. $1/6^{\text{th}}$ of the circumference   ii. $5/8^{\text{ths}}$ of the circumference

iii. One whole circumference   iii. One whole circumference   iii. One whole circumference

    d.   Would your answers to parts (a) through (c) change if the radii of the circles centered at the vertex of the respective angles were twice as long? Explain.

7. Use a piece of string and a compass to complete the following tasks.
    a. Use your compass to construct a circle that has a radius equal to the length of your piece of string. (*Members in your group should have strings of different lengths.*)

    b. Construct an angle whose vertex is at the center of your circle and that cuts off an arc that has the length of your piece of string (one radius length). Compare the openness of your angle with that of your classmates. What do you notice?

    c. Use your piece of string to create an angle that cuts off an arc that measures 2.5 radius lengths. Compare the openness of your angle with those of your classmates. What do you notice?

    d. Using your piece of string (the circle's radius) as a unit of measurement, how many radius lengths rotate along the circumference of your circle? Compare your answer with those of your classmates. What do you notice? Why did you get this result?

Radians

A **radian** is a unit of angle measure. An angle that measures 1 radian subtends an arc that is one radius length of any circle centered at its vertex. *Note that an arc that is 1 radius length is also $\frac{1}{2\pi}$ times as long as the circle's circumference.*

---

    e. Why is a radian a convenient unit for measuring angles?

    f. Does the radius of a circle centered at the vertex of an angle satisfy the criterion for a unit of angle measure discussed in Exercise #4? Explain.

8. Describe what it means for an angle to have the following angle measures and represent these angles on the given circle.
    a. 1 radian   b. 2 radians

    c. 3.5 radians

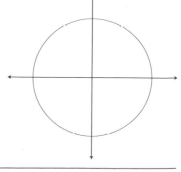

*9. Describe what it means for an angle to have a measure of
    a. 1 degree   b. 10 degrees   c. 47 degrees

Degrees

A ***degree*** is a unit of angle measure. An angle that measures 1 degree (or 1°) subtends an arc that is $\frac{1}{360}$ times as long as the circumference of any circle centered at its vertex.

---

10. a. If an angle has a measure of 1.5 radians, what percentage of the entire circle's circumference does that angle subtend?

    b. If an angle subtends an arc that is 23.87% of the entire circle's circumference, what is the measure of that angle in degrees?

11. Convert the following angle measures given in radians to measures in degrees.
    a. 2π radians   b. 2.7 radians   c. 4 radians   d. $\frac{\pi}{2}$ radians

12. Convert the following angle measures given in degrees to measures in radians.
    a. 52 degrees   b. 90 degrees   c 243 degrees

13. a. Define a function f that converts an angle measure in radians to its equivalent measure in degrees, and then define its inverse.

    b. Define f^{-1} and describe its input and output quantities.

    c. Explain how f and f^{-1} are related.

    d. Evaluate each of the following.

        i. $f(1)$ ii. $f^{-1}(120)$ iii. $f^{-1}\!\left(f\!\left(\frac{\pi}{2}\right)\right)$ iv. $f\!\left(f^{-1}(180)\right)$

---

1. Use a compass and a piece of string to estimate the measure of the given angle in radians. Explain what your estimate represents.

*2. a. Given that the following angle measures $\theta = 0.45$ radians, determine the length of each arc cut off by the angle. Assume the circles to have radii of 2 inches, 2.4 inches, and 2.9 inches. (*Drawing not to scale.*)

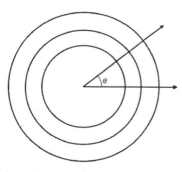

b. Consider any circle that is centered at the vertex of an angle measuring θ radians. Define a formula that relates the following three quantities.
- the arc length (in inches) cut off by the angle
- the angle measure (in radians)
- the circle's radius (in inches).   *Justify your formula.*

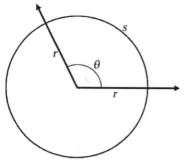

*3. What percentage of a circle's circumference is cut off by an angle that is 35 degrees plus 1.5 radians? Illustrate this angle measure below.

*4. A fellow student announced that the measure of the given angle is 1.2 inches. Is this an appropriate measure? Why or why not?

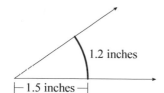

5. Use the given illustration to answer the following questions. (*Images not drawn to scale*).
a. A circle has a 12-inch radius. An angle, whose vertex is at the circle's center, cuts the circle into two parts and subtends the longer arc. The angle measures $\frac{3\pi}{2}$ radians. How long is the shorter arc (in inches)?

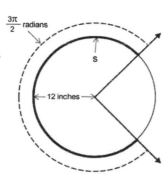

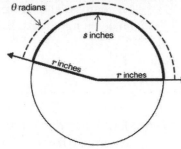

b. A circle has a radius of *r* inches. An angle with a vertex is at the circle's center cuts off an arc that is *s* inches long. Write a formula that conveys the relationship between the radius *r* (in inches), the angle measure θ (in radians), and the subtended arc length *s* (in inches).

c. A circle has a 4.8-inch radius. An angle with a vertex at the circle's center cuts the circle into two arcs, with one arc measuring 19.6 inches. What is the measure of this angle in radians?

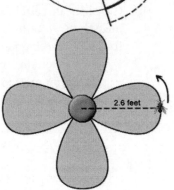

For Exercises #6-7, a bug sits on the end of a fan blade as the blade rotates in a counter-clockwise direction. The bug is 2.6 feet from the center of the fan and is located at the 3 o'clock position as the blade begins to turn.

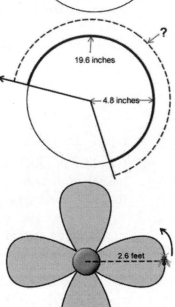

*6. a. If the bug travels 5.2 feet around the circumference of the circle from the 3 o'clock position, what angle measure (in radians) has the fan blade swept out?

b. If the bug travels 11.7 feet around the circumference of the circle from the 3 o'clock position, what angle measure (in radians) has the fan blade swept out?

c. Express the angle measure θ (in radians) swept out by the fan blade in terms of the arc length traveled by the bug *s* and the circle's radius *r* (both measured in feet).

d. If θ remains the same and the radius *r* of the fan doubles, how will the arc length *s* change?

7. How many feet has the bug traveled around the circle swept out by the tip of the fan blade when the angle swept out by the fan blade measures 0.765 radians? $\frac{\pi}{2}$ radians?

A bug sits on the end of a fan blade as the blade rotates in a counter-clockwise direction. The bug is 2.6 feet from the center of the fan and is located at the 3 o'clock position as the blade begins to turn.

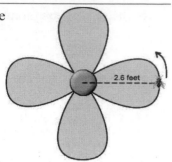

*1. What quantities might you track to represent the location of a bug as it sits on the tip of a rotating fan blade? List as many quantities as possible.

*2. a. Explain how the bug's "vertical" distance (as viewed in the diagram) above the horizontal diameter changes as the bug travels from:
   • the 3 o'clock position to the 12 o'clock position

   • the 12 o'clock position to the 9 o'clock position

   • the 9 o'clock position to the 6 o'clock position

   • the 6 o'clock position to the 3 o'clock position

   b. As the bug travels from the 3 o'clock position to the 12 o'clock position, how does the angle measure swept out by the fan blade change?

   c. For successive equal changes of angle measure from the 3 o'clock position to the 12 o'clock position, how do the corresponding *changes* in vertical distance change? Identify these changes on a diagram of the situation.

   d. Create a graph to illustrate how the bug's vertical distance above the horizontal diameter (measured in feet) co-varies with the measure of the angle θ swept out by the bug's fan blade (measured in radians). Plot points using the axes to illustrate how the angle and the vertical distance co-vary.

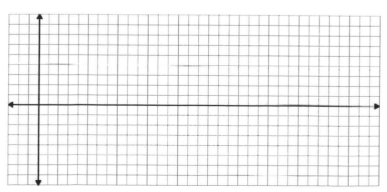

e.   How will your graph in (d) change if the radius of the fan is 3.9 feet instead of 2.6 feet?

f.   Create a graph to illustrate how the bug's vertical distance above the horizontal diameter (measured in radius lengths, or radii) co-varies with the measure of the angle θ swept out by the bug's fan blade (measured in radians). Plot points using the axes to illustrate how the angle and the vertical distance co-vary.

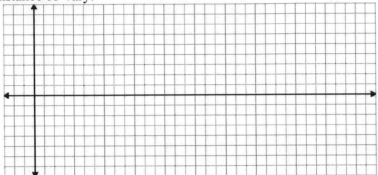

g.   Explain why the graphs you created in part (d) and (f) display the same general behavior.

The function you have been working with and graphing in Exercise #2 is called the *sine function.* It is given a special name because of how common the function is in math and science applications. The sine function is part of a family of functions called *trigonometric functions*.

The Sine Function

A circle is centered at the origin of a coordinate plane. An angle with its vertex at the circle's center has its initial ray pointing in the 3 o'clock position and its terminal ray rotates counterclockwise.

Call the point where the terminal ray intersects the circle the *terminal point.*

The function that represents the terminal point's vertical distance above the horizontal diameter (measured in radius lengths, or radii) in terms of the angle of rotation's measure is called the *sine function.* We often abbreviate sine as "sin" in mathematical expressions.

The angle measure θ is the input to the sine function and the output $\sin(\theta)$ is a directed distance above the horizontal diameter measured in radius lengths.

We say "directed distance" here because the value is considered to be negative if the distance is measured in the downward direction.

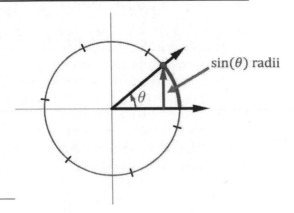

3. Label a point (x, y) on the circle, then label θ and $\sin(\theta)$ as defined previously.

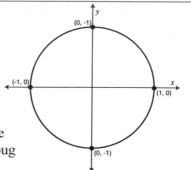

4. a. Explain how the bug's "horizontal" distance (as viewed in the diagram) to the right of the vertical diameter changes as the bug travels from:
   - the 3 o'clock position to the 12 o'clock position

   - the 12 o'clock position to the 9 o'clock position

   - the 9 o'clock position to the 6 o'clock position

   - the 6 o'clock position to the 3 o'clock position

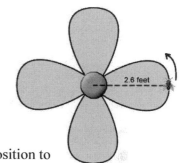

   b. For successive equal changes of angle measure from the 9 o'clock position to the 3 o'clock position, how do the corresponding *changes* in horizontal distance change? Identify these changes on a diagram of the situation.

   c. Create a graph to illustrate how the bug's *horizontal distance to the right of the vertical diameter* (measured in feet) co-varies with the measure of the angle θ swept out by the bug's fan blade (measured in radians). Plot points on the axes to illustrate how the angle and the horizontal distance co-vary.

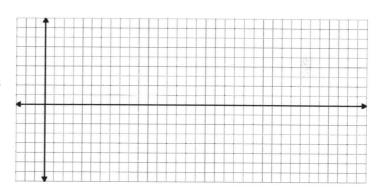

   d. How will your graph in (c) change if the radius of the fan is 4.2 feet instead of 2.6 feet?

   e. Create a graph to illustrate how the bug's horizontal distance to the right of the vertical diameter (measured in radius lengths) co-varies with the measure of the angle θ swept out by the bug's fan blade (measured in radians).

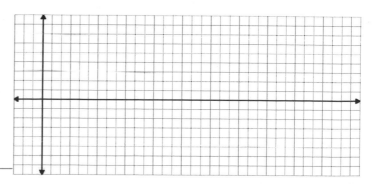

The function you have been working with and graphing in Exercise #4 is called the **cosine function**.

The Cosine Function

A circle is centered at the origin of a coordinate plane. An angle with its vertex at the circle's center has its initial ray pointing in the 3 o'clock position and its terminal ray rotates counterclockwise.

The function that represents the terminal point's horizontal distance to the right of the vertical axis (measured in radius lengths, or radii) in terms of the angle of rotation's measure is called the **cosine function**. We often abbreviate cosine as "cos" in mathematical expressions.

The angle measure θ is the input to the cosine function and the output $\cos(\theta)$ is a directed distance to the right of the vertical diameter measured in radius lengths.

We say "directed distance" here because the value is considered to be negative if the distance is measured to the left.

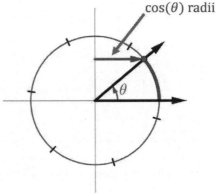

5. Label a point (x, y) on the unit circle, then label θ and $\cos(\theta)$ as defined.

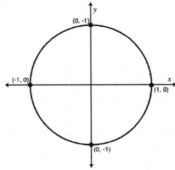

*6. Use this figure to answer the questions.

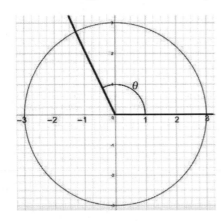

    a. Approximate the measure of θ in radians.

    b. Without a graphing calculator, approximate the value of $3 \cdot \sin(\theta)$.

    c. Using your answer from part (b), approximate the value of $\sin(\theta)$.

    d. Without a graphing calculator, approximate the value of $3 \cdot \cos(\theta)$.

    e. Using your answer from part (d), approximate the value of $\cos(\theta)$.

Note: Consider all values of θ in this investigation to be expressed in radians.

An arctic village skiing trail has a radius 1 kilometer long. A skier started at the position (1, 0) on the coordinate axes and skied counter-clockwise.

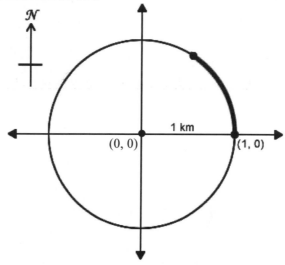

1. a. Label the points (0, 1), (−1, 0) and (0, −1) on the coordinates axes in the diagram.

    b. What is the angle measure (in radians) of the angles that have an initial ray passing through (0, 0) and (1, 0) and a second ray that passes through (0, 0) and the point:
        i. (0, 1)   ii. (0, −1)   iii. (−1, 0)

    c. Illustrate the radian measure (in decimal form) of each of the angles described in part (b).

*2. An arctic village skiing trail has a 1-kilometer radius. A skier started at the position (1, 0) on the coordinate axes and skied counter-clockwise. After skiing counter-clockwise, the skier paused for a rest at the point (0.5, 0.87).
    a. Describe what the value 0.5 represents and illustrate this measurement (value) on the axes from Exercise #1.

    b. Describe what the value 0.87 represents and illustrate this measurement (value) on the axes from Exercise #1.

3. Recall that the sine and cosine functions are defined as
$x = \cos(\theta)$ and $y = \sin(\theta)$, given that (x, y) is a point on
the unit circle and θ is the measure of the angle that
subtends the arc between $(1, 0)$ and (x, y).

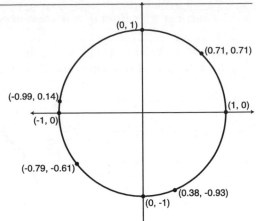

a. Use the unit circle to generate a graph of $y = \sin(\theta)$
where θ is the measure of the angle subtending the
arc between $(1, 0)$ and (x, y) rotated. *Label values on
your axes and 8 points on your graph using the
given points on the circle.*

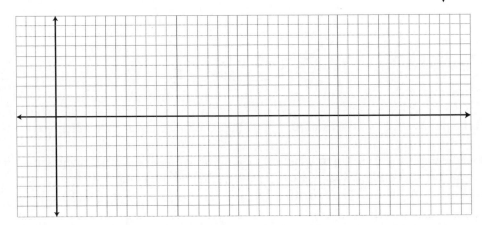

b. What is the domain of the sine function? What is the range of the sine function? As θ
varies from $\frac{\pi}{2}$ to π, how does the output of the sine function vary?

c. What is the concavity of the sine function on the interval, $\frac{\pi}{2} < \theta < \pi$? Provide a rationale for your
answer.

d. Use the unit circle above to generate a graph of $x = \cos(\theta)$. *Note that in this case, values of x are
tracked on the vertical axis and values of θ are tracked on the horizontal axis.*

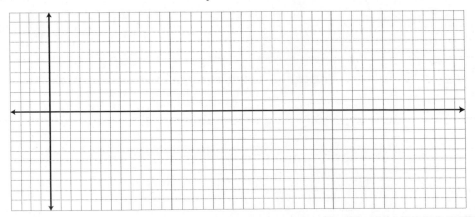

e. As θ varies from 0 to $\frac{\pi}{2}$, how does $x = \cos(\theta)$ vary?

*4. A second arctic village maintains a circular cross-country ski trail that has a 2.5-kilometer radius.

a. A skier started skiing from the position (2.5, 0) and skied counter-clockwise for 2.75 kilometers before stopping for a rest.

i. Explain what $\frac{2.75}{2.5}$ represents in this context.

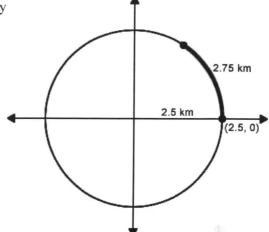

ii. Determine the ordered pair (measured in radius lengths) on the coordinate axes that identifies the location where the skier rested.

iii. Determine the ordered pair (measured in kilometers) on the coordinate axes that identifies the location where the skier rested.

b. The skier then continued along the trail, traveling counter-clockwise until he had traveled for a *total* of 10 kilometers before stopping for another rest.

i. Determine the ordered pair (measured in radius lengths) on the coordinate axes that identifies this location where the skier rested.

ii. Determine the ordered pair (measured in kilometers) on the coordinate axes that identifies this location where the skier rested.

c. Suppose that a second skier started skiing from the position (2.5, 0) and skied *clockwise* for 5 kilometers before stopping for a rest.

i. Determine the ordered pair (measured in radius lengths) on the coordinate axes that identifies this location where the skier rested.

ii. Determine the ordered pair (measured in kilometers) on the coordinate axes that identifies this location where the skier rested.

5. Consider a coordinate system where the coordinate values of a point are measured as a number of radius lengths. What is the general form of an ordered pair (expressed in a number of radius lengths) of any point on a circle that has a radius of r kilometers and an angle measure of θ radians?

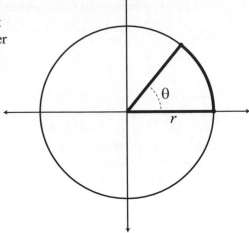

*6. a.   Construct angles on the unit circle with these radian measures: $\frac{\pi}{6}, \frac{5\pi}{6}, \frac{7\pi}{6}, \frac{11\pi}{6}$.

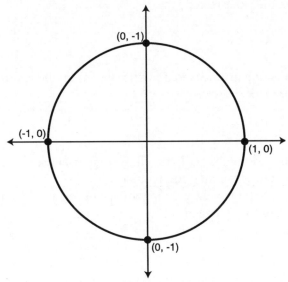

b. Using your calculator, determine the coordinates corresponding to the angle with a measure of $\frac{\pi}{6}$ radians. Describe what these coordinates represent.

c. Using the coordinates determined in part (b), determine the coordinates that correspond to $\frac{5\pi}{6}, \frac{7\pi}{6}, \frac{11\pi}{6}$ radians. Explain how you determined these coordinates. *Do not use a calculator.*

José extends his arm straight out, holding a 2.3-foot string with a ball on the end. He twirls the ball around in a circle with his hand at the center so that the plane in which it is twirling is perpendicular to the ground. Assume that the ball twirls counter-clockwise starting at the 3 o'clock position.

1. Label quantities on the diagram that could be used to describe the position of the ball as it rotates around José's hand.

*2. Define a formula for function f that relates the ball's *vertical distance above* Jose's hand (in feet) as a function of the angle of rotation's measure (in radians) as the ball twirls counter-clockwise from a 3 o'clock position. *Be sure to define any variables you use.*

*3. Jose twirls the ball so that it travels one radius length per second along its circular path.
    a.   What is the measure of the angle of rotation θ after t seconds?

    b.   Define a function g that relates the ball's *vertical distance above* José's hand as a function of the number of seconds elapsed. *Be sure to define any variables you use.*

    c.   Over what time interval does the ball complete one revolution?

*4. Jose twirls the ball so that it travels two radius lengths per second along its circular path.
    a.   What is the measure of the angle of rotation θ after t seconds?

b.   Define a function h that relates the ball's *vertical distance above* Jose's hand as a function of the number of seconds elapsed. *Be sure to define any variables you use.*

c.   Over what time interval does the ball complete one revolution?

*5. Jose twirls the ball so that it travels 0.5 radius lengths per second along its circular path.
   a.   What is the measure of the angle of rotation θ after t seconds?

b.   Define a function j that relates the ball's *vertical distance above* José's hand as a function of the number of seconds elapsed. *Be sure to define any variables you use.*

c.   Over what time interval does the ball complete one revolution?

*6. Graph the functions determined in Exercises #3-5. On each graph illustrate an interval of input values over which the function values complete one full cycle (i.e., an interval of t for which the ball completes one revolution).
   a.   Complete the table of values and then graph g (see Exercise #3). Label your axes.

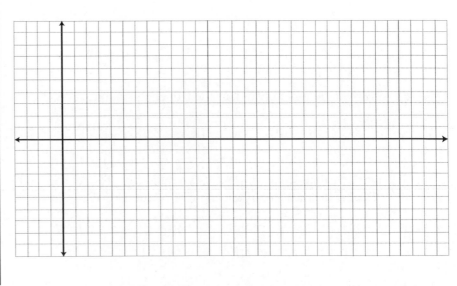

$g(t) =$	
t	$g(t)$
0	
$\frac{\pi}{2}$	
π	
$\frac{3\pi}{2}$	
2π	
$\frac{5\pi}{2}$	
3π	
$\frac{7\pi}{2}$	
4π	

b. Complete the table of values and then graph *h* (see Exercise #4). Label your axes.

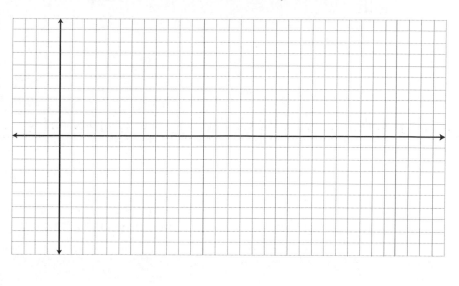

$h(t) =$	
t	*h(t)*
0	
$\frac{\pi}{4}$	
$\frac{\pi}{2}$	
$\frac{3\pi}{4}$	
π	
$\frac{5\pi}{4}$	
$\frac{3\pi}{2}$	
$\frac{7\pi}{4}$	
2π	

c. Complete the table of values and then graph *j* (see Exercise #5). Label your axes.

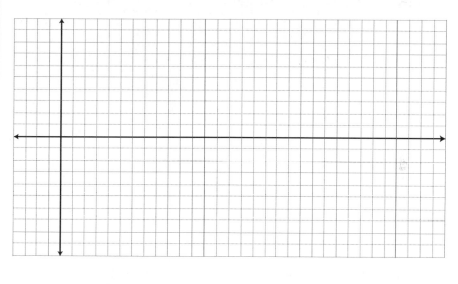

$j(t) =$	
t	*j(t)*
0	
$\frac{\pi}{2}$	
π	
$\frac{3\pi}{2}$	
2π	
$\frac{5\pi}{2}$	
3π	
$\frac{7\pi}{2}$	
4π	

The Period of a Function

The **period** of a function is the smallest interval of input values needed to complete one full cycle of output values. Not every function has a period. Functions with a period are called **periodic functions**.

The Amplitude of a Periodic Function

The **amplitude** of a periodic function represents half the difference between the function's maximum and minimum values. In the context of trigonometric functions it can be interpreted to be the radius of the circle measured in some length unit.

---

d. Describe the reasoning you used to determine the period of the functions in (a), (b), and (c) above.

e. What is the amplitude of the functions in parts (a) through (c)?

7. Jose twirls the ball so that it travels 3.8 radius lengths per second along its circular path.
   a. What is the measure of the angle of rotation θ after t seconds?

   b. Define a function k that relates the ball's *vertical distance above* José's hand as a function of the number of seconds elapsed. *Be sure to define any variables you use.*

   c. Over what time interval does the ball complete one revolution?

   d. What is the amplitude for function k?

8. Suppose you determine that it takes the ball 20 seconds to complete one revolution.
   a. Determine the speed of the ball in radius lengths per second along its circular path.

   b. Define a function m that relates the ball's *vertical distance above* José's hand as a function of the number of seconds elapsed. *Be sure to define any variables you use.*

*1. A Ferris wheel has a 52-foot radius and the horizontal diameter is located 58 feet off the ground.

   a. If the Ferris wheel rotates at a constant rate of ¼ radian per minute, how many minutes does it take for the Ferris wheel to make one full revolution?

   b. Define a function g to represent the distance (in feet) of a Ferris wheel bucket above the *ground* as a function of the number of seconds, t, since the Ferris wheel began rotating from the 3 o'clock position. Assume the Ferris wheel rotates at a constant rate of ¼ radian per minute.

   c. Graph the function in part (b) using a calculator or graphing software. Compare the graph to the graph of $f(t) = \sin(t)$ and explain the reason for any differences.

   d. How would the function definition and graph change if the Ferris wheel's radius is 45 feet instead of 52 feet long?

   e. Suppose the Ferris wheel rotates at a constant rate of 3 radians per minute instead. How many minutes will it take for the Ferris wheel to make one full revolution?

   f. What might the function $h(t) = \sin(2t)$ represent in this situation? (*Hint: Consider the meaning of the value of 2t.*)

2. a. What is the period of the function f, defined by $f(\theta) = \sin(\theta)$? What does the period of f represent?

   b. What is the amplitude of f ? Why?

3. Let $j(\theta) = \sin\left(\frac{\theta}{2}\right)$.

   a. By how much must θ vary in order for $\frac{\theta}{2}$ to vary by 2π radians? What does this tell us?

   b. Let $f(\theta) = \sin(\theta)$, $j(\theta) = \sin\left(\frac{\theta}{2}\right)$, and $p(\theta) = 3\sin\left(\frac{\theta}{2}\right)$. Graph all three functions on the given axes. *Label the axes before constructing your graphs.*

   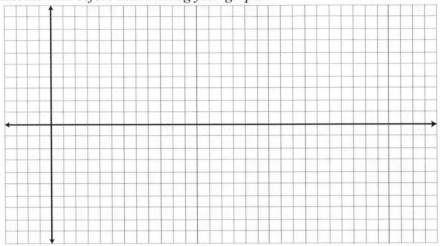

   c. Demonstrate the period and amplitude of each function on its graph.

*4. The function h is defined by $h(\theta) = \sin(4\theta) + 2$ where θ is an angle measure (in radians).

   a. Complete the table of values.

θ	$h(\theta)$
0	
$\frac{\pi}{8}$	
$\frac{\pi}{4}$	
$\frac{3\pi}{8}$	
$\frac{\pi}{2}$	
$\frac{5\pi}{8}$	
$\frac{3\pi}{4}$	

   b. Graph h. *Be sure to label your axes.*

   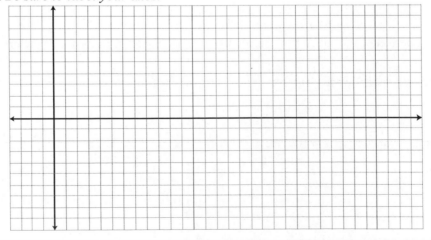

c.   What is the period of *h*? What does the period of *h* represent?

d.   What is the amplitude of *h*?

*5. A water wheel (used for power in the 18th century) is illustrated to the right.

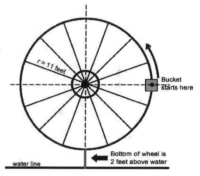

a.   Sketch a graph of the *height* of the bucket above the water (in feet) as a
function of the angle of rotation (measured in radians) of the bucket from
the 3 o'clock position. *Label your axes*.

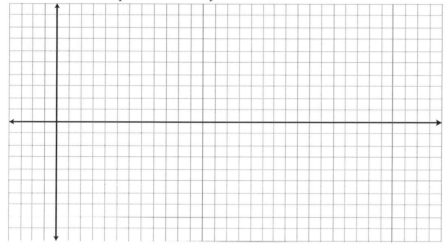

b.   Define a function *f* that represents the bucket's distance *d* above the water (in feet) as a function
of the angle of rotation (measured in radians) of the bucket from the 3 o'clock position. *Be sure to
define your variables.*

c.   Define a function *g* that represents the bucket's distance *d* above the water (in feet) as a function
of the *number of feet* the bucket has rotated counter-clockwise from the 3 o'clock position. *Be
sure to define your variables.*

*6. The graph of function *h* is given.

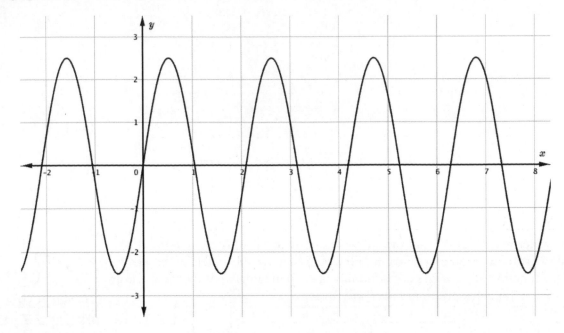

a. What is the approximate value of the period of *h*? (Try to estimate its value in terms of π.)

b. What is the approximate value of the amplitude of *h*?

c. Define a formula that could represent function *h*.

7. Let $k(\theta) = 2\sin\left(\frac{3}{7}\theta\right) + 5$.
    a. Determine the period of *k*.

    b. Determine the amplitude of *k*.

    c. How might graphing the line $y = 5$ help you construct the graph of *k*?

*1. Consider function f shown in the given graph.

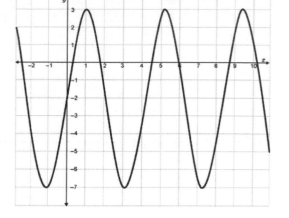

  a. What is its period?

  b. What is its minimum value?

  c. What is its maximum value?

  d. What is the equation of the horizontal line that falls halfway between the function's maximum and minimum values? (This is called the function's *midline*.)

  e. What is its amplitude?

  f. Define a formula for function f .

*2. Let's return to thinking about a Ferris wheel. We've already considered how the height of the Ferris wheel's center affects the height of a rider as the wheel rotates. For example, a certain Ferris wheel has its center 30 feet off the ground with a 25-foot radius. Shawna is at the 3 o'clock position when the Ferris wheel begins to rotate. *Note that it takes 20 seconds for the Ferris wheel to make one full rotation.*

  a. Draw a diagram of this situation.

  b. Define a formula for the function relationship in the given graph and explain what the graph represents in terms of the quantities in the situation. *For example, what two quantities are being coordinated? What information from the context influences the function's amplitude, period, etc.?*

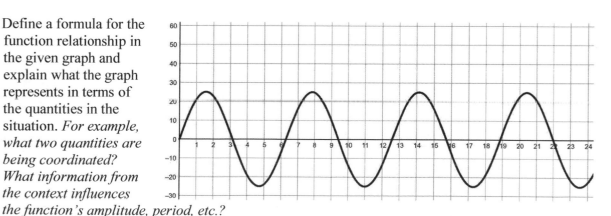

c. Define a formula for the function relationship in the given graph and explain what the graph represents in terms of the quantities in the situation. *For example, what two quantities are being coordinated? What information from the context influences the function's amplitude, period, etc.?*

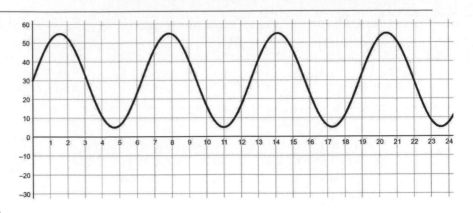

d. Compare and contrast the functions in parts (a) and (b). How are they different and why?

e. Define a formula for the function relationship in the given graph and explain what the graph represents in terms of the quantities in the situation. *For example, what two quantities are being coordinated? What information from the context influences the function's amplitude, period, etc.?*

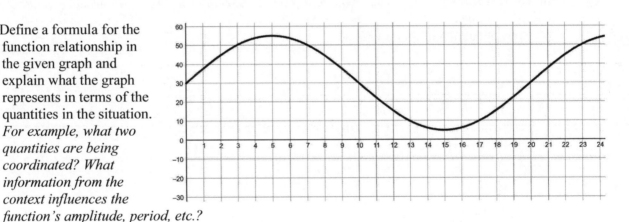

Let's consider a different scenario with the Ferris wheel. Since the mathematical convention is to measure angles from the 3 o'clock position, we have typically had riders on the Ferris wheel starting at the 3 o'clock position. But what if they start from a different position?

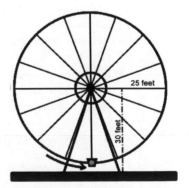

*3. Shawna boards the Ferris wheel shown from the bottom (the 6 o'clock position) and the Ferris wheel rotates counter-clockwise.
  a. After the Ferris wheel rotates π radians (half of one full rotation), what is her height off the ground?

  b. Your classmate says that the expression $25\sin(\pi) + 30$ should calculate Shawna's height above the ground (in feet) after the Ferris wheel rotates π radians. Do you agree? Explain.

c. After the Ferris wheel rotates through each of the following rotation angles, what is Shawna's height above the ground? *Write an expression that represents her height using the sine function and then provide a decimal approximation of her height above the ground (in feet). We recommend drawing a diagram of each situation first.*

  i. $\frac{\pi}{2}$ radians

  ii. 2π radians

  iii. $\frac{3\pi}{4}$ radians

  iv. $\frac{5\pi}{4}$ radians

  v. $\frac{3\pi}{2}$ radians

  vi. $\frac{\pi}{4}$ radians

d. Sketch a graph of the function f that represents Shawna's height above the ground (in feet) in terms of the measure of the rotation angle since she boarded (θ, in radians).

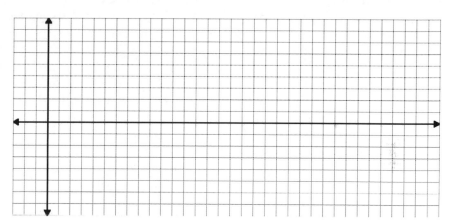

e. Define the formula for function f.

f. Define the formula for function g that represents Shawna's horizontal distance to the right of the Ferris wheel's vertical diameter in terms of the measure of the rotation angle since she boarded (θ, in radians).

g. Graph g using a graphing calculator or graphing software and discuss with a partner or as a class how the graph supports your understanding of the function's behavior.

*4. Using the scenario in Exercise #3, complete the following.
   a. If the Ferris wheel is rotating at 0.1 radians per second, define the formula for a function that represents Shawna's height above the ground (in feet) in terms of the number of seconds since the Ferris wheel started to rotate.

   b. Repeat part (a) if the rotation speed is instead 0.5 radians per second.

   c. Repeat part (a) if the rotation speed is instead 2 radians per second (*even though that's dangerously fast!*).

   d. What impact does the rotation speed have on the function and its graph? (For example, does the rotation speed impact the amplitude? The midline? Something else?)

5. Shawna instead starts at the top of the Ferris wheel (the 12 o'clock position) and the Ferris wheel rotates counter-clockwise.
   a. After the Ferris wheel rotates through each of the following rotation angles, what is Shawna's height above the ground? *Write an expression that represents her height using the sine function and then provide a decimal approximation of her height above the ground (in feet). We recommend drawing a diagram of each situation first.*
      i. $\frac{\pi}{2}$ radians   ii. π radians   iii. $\frac{\pi}{3}$ radians

   b. Define the formula for function f that represents Shawna's height above the ground (in feet) in terms of the measure of the rotation angle since she boarded (θ, in radians).

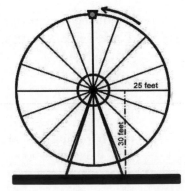

25 feet

30 feet

   c. Graph f using a graphing calculator or graphing software and discuss with a partner or as a class how the graph supports your understanding of the function's behavior.

d. If the Ferris wheel is rotating so that it completes a rotation in 50 seconds, define the formula for a function that represents Shawna's height above the ground (in feet) in terms of the number of seconds since the Ferris wheel started to rotate.

6. Shawna instead starts at the position shown ($\frac{1}{8}$ of a rotation counter-clockwise from the 3 o'clock position) and the Ferris wheel rotates counter-clockwise.
    a. After the Ferris wheel rotates through each of the following rotation angles, what is Shawna's height above the ground? *Write an expression that represents her height using the sine function and then provide a decimal approximation of her height above the ground (in feet). We recommend drawing a diagram of each situation first.*
        i. $\frac{\pi}{2}$ radians   ii. 2π radians   iii. $\frac{\pi}{6}$ radians

    b. Define the formula for function f that represents Shawna's height above the ground (in feet) in terms of the measure of the rotation angle since she boarded (θ, in radians).

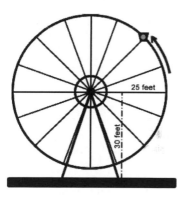

    c. Graph f using a graphing calculator or graphing software and discuss with a partner or as a class how the graph supports your understanding of the function's behavior.

    d. If the Ferris wheel is rotating at 0.12 radians per second, define the formula for a function that represents Shawna's height above the ground (in feet) in terms of the number of seconds since the Ferris wheel started to rotate.

7. Predict how the graphs of the two functions will be similar and different. Then graph each pair of functions to test your prediction.

   *a. $f(\theta) = \sin(\theta)$ and $g(\theta) = 12.5\sin(\theta)$

   *b. $h(\theta) = 12.5\cos(\theta)$ and $j(\theta) = 12.5\cos\left(\theta - \frac{\pi}{3}\right)$

   c. $p(\theta) = 12.5\sin(\theta)$ and $q(\theta) = 12.5\sin(2\theta)$

   *d. $r(\theta) = \cos(\theta)$ and $s(\theta) = \cos\left(\frac{\theta}{3}\right)$

   e. $v(\theta) = \sin(\theta)$ and $w(\theta) = \sin\left(\theta + \frac{\pi}{2}\right)$

1. So far in this module we've primarily been working with circles where the radius is known. But what if that isn't the case?

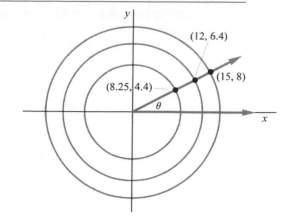

   a. Come up with a strategy for determining the angle measure θ (in radians).

   b. What do you notice about the ratio of the y and x coordinates for each coordinate shown in the diagram?

   c. If we continued drawing more and more circles centered at the angle's vertex, list at least three more points that could fall at the intersection of the angle's terminal ray and one of these circles.

2. Consider the graph below, which shows the first quadrant of the coordinate plane [the bottom left corner is (0, 0), A is located at (1,0), and J is located at (0,1)]. Assume the distance between consecutive points on the curve represents $1/9^{th}$ of a quarter circle. The circle's radius is 1 unit.

   *a. Consider the angle formed by the initial ray (passing through the origin and point A) and the terminal ray (passing through origin and point B).

   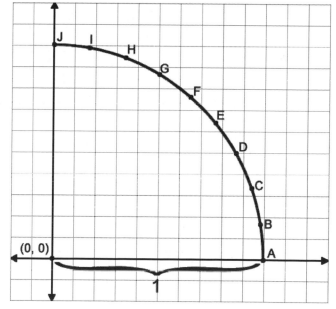

      i. What is the measure of this angle in radians?

      ii. What is the ordered pair (x, y) describing the location of point B? [*First estimate the coordinates from the graph, then determine how to find more exact coordinates using a calculator.*]

   iii. What is the slope of the terminal ray (the ray passing through the origin and point B)?

b. Consider the angle that is formed by the initial ray (passing through the origin and point A) and the terminal ray (passing through origin and point C).

   i.   What is the measure of this angle in radians?

   ii. What is the ordered pair (x, y) describing the location of point C? [*First estimate the coordinates from the graph, then determine how to find more exact coordinates using a calculator.*]

   iii. What is the slope of the terminal ray (the ray passing through the origin and point C)?

*c. Complete the table below.

Point	Measure of the angle formed by the initial ray and the terminal ray passing through the given point (in radians), θ	Ordered pair describing the point's location (x, y)	Slope of the terminal ray passing through the given point, m
A	0	(1,0)	
B	$\frac{\pi}{18}$	about (0.985, 0.174)	
C	$\frac{2\pi}{18}$ or $\frac{\pi}{9}$	about (0.940, 0.342)	
D		about (0.866, 0.5)	about 0.577
E	$\frac{4\pi}{18}$ or $\frac{2\pi}{9}$	about (0.766, 0.643)	about 0.839
F	$\frac{5\pi}{18}$	about (0.643 0.766)	about 1.192
G	$\frac{6\pi}{18}$ or $\frac{\pi}{3}$		
H	$\frac{7\pi}{18}$	about (0.342, 0.940)	about 2.747
I	$\frac{8\pi}{18}$ or $\frac{4\pi}{9}$	about (0.174, 0.985)	
J	$\frac{9\pi}{18}$ or $\frac{\pi}{2}$	(0,1)	

*d. What happens to the slope of the terminal ray as the value of θ approaches $\frac{\pi}{2}$?

*e. For what angle measure θ (between $\theta = 0$ and $\theta = \frac{\pi}{2}$) will the slope of the terminal ray be 1?

*f. How does the slope of the terminal ray for a given angle measure relate the sine and cosine values for the same angle measure?

*3. Now let's think about how the slope of the terminal ray and the angle measure co-vary in the second quadrant (that is, when $\frac{\pi}{2} < \theta < \pi$).

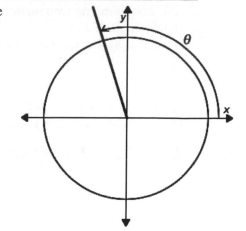

a. When $\frac{\pi}{2} < \theta < \pi$, is the slope of the terminal ray positive or negative? Justify your answer.

b. As θ increases from $\theta = \frac{\pi}{2}$ to $\theta = \pi$, how does the slope of the terminal ray vary?

c. For what angle measure θ (between $\theta = \frac{\pi}{2}$ and $\theta = \pi$) will the slope of the terminal ray be -1?

*4. Now let's think about how the slope of the terminal ray and the angle measure co-vary in the third quadrant (that is, when $\pi < \theta < \frac{3\pi}{2}$).

a. When $\pi < \theta < \frac{3\pi}{2}$, is the slope of the terminal ray positive or negative? Justify your answer.

b. As θ increases from $\theta = \pi$ to $\theta = \frac{3\pi}{2}$, how does the slope of the terminal ray vary?

*5. Now let's think about how the slope of the terminal ray and the angle measure co-vary in the fourth quadrant (that is, when $\frac{3\pi}{2} < \theta < 2\pi$).

a. When $\frac{3\pi}{2} < \theta < 2\pi$, is the slope of the terminal ray positive or negative? Justify your answer.

b. As θ increases from $\theta = \frac{3\pi}{2}$ to $\theta = 2\pi$, how does the slope of the terminal ray vary?

6. For what angle measure(s) θ (between $\theta = \pi$ and $\theta = 2\pi$) will the slope of the terminal ray be
   a. 1?   b. -1?   c. greater than 1?   d. less than -1?

The function you've been working with is called the ***tangent function***. It is another periodic trigonometric function.

> ### The Tangent Function
>
> The tangent function $f(\theta) = \tan(\theta)$ inputs an angle measure (counter-clockwise from the 3 o'clock position) and outputs the slope of the angle's terminal ray. In addition, this value represents the relative size of the sine and cosine values for the same angle measure.
>
> *Note that since the terminal ray is sometimes vertical, there are angle measures for which the tangent function is undefined.*

*7. Without using a calculator determine if each of the following statements is true or false. Justify your answer by drawing a diagram or with a written explanation.

a. $\tan\left(\frac{2\pi}{3}\right) < 0$

b. $\tan\left(\frac{\pi}{5}\right) < 0$

c. $0 < \tan\left(\frac{4\pi}{3}\right)$

d. $\tan\left(\frac{7\pi}{8}\right) > 0$

8. Use a calculator to evaluate each of the following and then interpret the meaning of your answers as i) the slope of a terminal ray and ii) the relative size of a sine and cosine value.

a. $\tan(75°)$

b. $\tan\left(\frac{\pi}{7}\right)$

c. $\tan\left(\frac{3\pi}{5}\right)$

9. Sketch a graph of the slope of the terminal ray versus the angle measure as the angle measure varies from 0 radians to 2π radians.

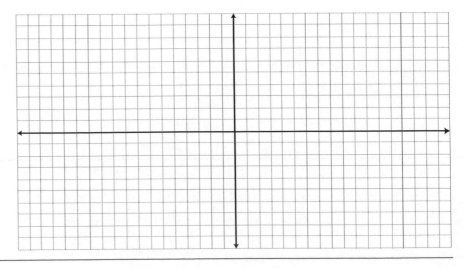

10. The tangent function is periodic because it repeats at regular intervals.
    a. What is the period of the tangent function?

    b. Where are the vertical asymptotes located and why do they occur for these values of θ ?

    c. Use the following diagrams to help you explain why the slopes of the terminal ray are the same whenever the angle measure differs by π radians (or 180°).

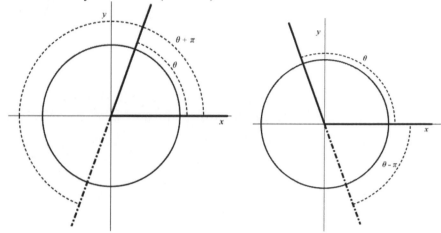

11. Guy wires are used to stabilize very tall, thin objects such as telephone poles and radio towers. A certain guy wire attached to a tower makes a 0.92-radian angle with the ground. *Diagram not drawn to scale.*

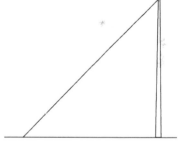

    a. If you walk from the point where the wire is attached to the ground and travel 20 feet toward the tower, how high off the ground is the guy wire at that location?

    b. If you walk from the point where the wire is attached to the ground and walk ANY distance (in feet) towards the tower, how high off the ground is the guy wire at that location?

    c. What information do you need to determine the height on the tower where the wire attaches?

*12. In Exercise #2 you were asked to calculate the slope of the terminal ray of the angle when you knew that the ray began at the origin and passed through a point on the circle. For example, if the terminal ray passed through the point (0.1736, 0.9848) then the slope is calculated as follows.

$$m = \frac{0.9848 - 0}{0.1736 - 0}$$
$$= \frac{0.9848}{0.1736}$$
$$\approx 5.673$$

a. Suppose you know that an angle measures θ radians using a circle with a radius of 5 units. How can you find the point (x, y) on the circle through which the terminal ray passes?

b. Using your answer to part (a), complete the following statement by writing $\tan(\theta)$ in terms of $\cos(\theta)$ and $\sin(\theta)$. (In other words, when we know the value of $\cos(\theta)$ and $\sin(\theta)$, how can we determine the value of $\tan(\theta)$?)

$\tan(\theta) =$

13. In earlier investigations we determined that we can find a point (x, y) on a circle using $x = r \cdot \cos(\theta)$ and $y = r \cdot \sin(\theta)$ where r is the circle's radius. If r is a circle's radius (in some unit), is the slope of the terminal side represented by $m = r \cdot \tan(\theta)$ or $m = \tan(\theta)$? Justify your answer.

Consider the bug on the end of the fan blade from Investigation 2. If we reverse the polarity in the fan motor, the blade will instead move in the clockwise direction. In order to distinguish the two possible rotation directions, mathematicians agreed upon the following conventions.

- Angles rotated counter-clockwise from the 3 o'clock position have positive measurements.
- Angles rotated clockwise from the 3 o'clock position have negative measurements.

This makes sense, since to undo a turn in the counter-clockwise direction, we must subtract, or turn the same amount clockwise.

*1. a. i. The fan rotates one-half of a complete rotation in the clockwise direction from the 3 o'clock position. What measurement do we use to describe the angle the fan blade sweeps out?

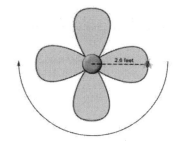

   ii. If the fan had rotated in the counter-clockwise direction instead, what angle of rotation is needed to end in the same position?

   b. i. The fan rotates three-quarters of a complete rotation in the clockwise direction from the 3 o'clock position. What measurement do we use to describe the angle the fan blade sweeps out?

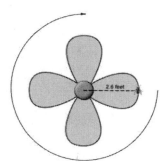

   ii. If the fan had rotated in the counter-clockwise direction instead, what angle of rotation is needed to end in the same position?

   c. Angles with terminal rays that coincide are called ***co-terminal angles***. Complete the following. Assume all angles are measured from the 3 o'clock position in radians.
   i. An angle measures α radians where α is a negative number. The angle is coterminal with an angle measuring $\frac{3\pi}{4}$ radians. What could be the value of α?

   ii. An angle measures α radians where α is a negative number. The angle is coterminal with an angle measuring $\frac{7\pi}{6}$ radians. What could be the value of α?

   iii. Define a function f that outputs the (negative) measure of an angle co-terminal with an angle measuring θ radians (with $0 \le \theta < 2\pi$).

*2. a. Fill in the table of values.

t	$\cos(t)$	$\cos(-t)$
0		
$-\frac{\pi}{6}$		
$-\frac{\pi}{4}$		
$-\frac{\pi}{3}$		
$-\frac{\pi}{2}$		
$-\frac{2\pi}{3}$		
$-\frac{3\pi}{4}$		
$-\frac{5\pi}{6}$		
$-\pi$		

b. How do the values of $\cos(t)$ and $\cos(-t)$ compare? (*Functions with this property are called **even functions**.*)

*3. a. Fill in the table of values.

t	$\sin(t)$	$\sin(-t)$
0		
$-\frac{\pi}{6}$		
$-\frac{\pi}{4}$		
$-\frac{\pi}{3}$		
$-\frac{\pi}{2}$		
$-\frac{2\pi}{3}$		
$-\frac{3\pi}{4}$		
$-\frac{5\pi}{6}$		
$-\pi$		

b. How do the values of $\sin(t)$ and $\sin(-t)$ compare? (*Functions with this property are called **odd functions**.*)

*4. Consider the bug on the end of the fan blade from Investigation 2. Suppose the fan has spun in the positive direction and swept out an angle measuring $\frac{9\pi}{4}$ radians.

a. What is the height of the bug (in feet) above the horizontal diameter of the fan?

b. What is the distance of the bug (in feet) to the right of the vertical diameter of the fan?

c. Provide the measures of two angles co-terminal with an angle measuring $\frac{9\pi}{4}$ radians.

5. Now suppose the fan spun clockwise so that we would describe the angle of rotation's measure as $-\frac{53\pi}{4}$ radians.

a. What is the height of the bug (in feet) above the horizontal diameter of the fan?

b. What is the distance of the bug (in feet) to the right of the vertical diameter of the fan?

c. Provide the measures of two angles co-terminal with an angle measuring $-\frac{53\pi}{4}$ radians.

6. a. Recall the Ferris wheel context from Investigation 6. A rider is initially sitting in a bucket at the 3 o'clock position. If the Ferris wheel rotates clockwise, what angle of rotation is necessary for the rider to reach the highest point on the Ferris wheel?

    b. What is the fewest number of radians the Ferris wheel must sweep out in either direction to return the rider to their original position?

Coordinating Related Sets of Measurements

Sometimes in mathematics we want to describe sets of numbers such as "every real number that is 0.1 greater than an integer". Numbers that satisfy this criterion include 1.1, 2.1, –0.9, and –1.9. But we could never list <u>all</u> of the numbers that satisfy this criterion individually.

Using mathematical notation, however, we could say "For every integer k, the set of numbers k + 0.1 satisfies the criterion".

Another example is the set of all even numbers. We can represent this as the set of $2n$ for all integers n.

---

*7. a. Represent the list of all multiples of $\frac{1}{4}$.

    b. Represent the list of all multiples of $\frac{\pi}{2}$.

    c. Represent the list of all real numbers $\frac{\pi}{3}$ units greater than a multiple of 2π.

---

*8. a. Write the list of angle measures (in radians) for angles co-terminal with an angle measuring 0 radians (measured counter-clockwise from the 3 o'clock position).

   b. Write the list of angle measures (in radians) for angles co-terminal with an angle measuring $\frac{\pi}{3}$ radians (measured counter-clockwise from the 3 o'clock position).

*9. a. Determine the values of x, $0 \le x < 2\pi$, such that $\cos(x) = \frac{\sqrt{3}}{2}$.

   b. Write the list of all values of x such that $\cos(x) = \frac{\sqrt{3}}{2}$.

10. a. Determine the values of x, $0 \le x < 2\pi$, such that $\sin(x) = \pm\frac{\sqrt{2}}{2}$.

   b. Write the list of all values of x such that $\sin(x) = \pm\frac{\sqrt{2}}{2}$.

*1. a. Complete the function machine diagrams by identifying the input and output quantities for each function and the domain and range of each function.

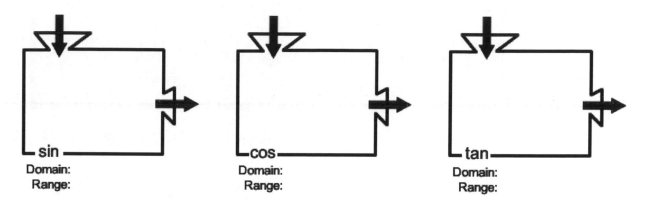

sin
Domain:
Range:

cos
Domain:
Range:

tan
Domain:
Range:

b. What is the input quantity for the inverse relation of the sine function?

c. What is the output quantity for the inverse relation of the sine function?

d. Is the inverse relation of the sine function a function itself (i.e., does each vertical distance above the horizontal diameter produce a unique angle measure)? Explain.

The Inverse Sine Function

When considering the ***inverse sine function*** (or the ***arcsine function***) we first restrict the domain of the sine function to $-\frac{\pi}{2} \le \theta \le \frac{\pi}{2}$.

The inverse sine function takes as its input the vertical distance of the terminal point above the horizontal diameter (a sine value) and outputs a single angle measure that produces that sine value.

We write this function as $\theta = \arcsin(y)$ or $\theta = \sin^{-1}(y)$.

The Inverse Cosine Function

When considering the ***inverse cosine function*** (or the ***arccosine function***) we first restrict the domain of the cosine function to $0 \le \theta \le \pi$.

The inverse cosine function takes as its input the horizontal distance of the terminal point to the right of the vertical diameter (a cosine value) and outputs a single angle measure that produces that cosine value.

We write this function as $\theta = \arccos(x)$ or $\theta = \cos^{-1}(x)$.

The Inverse Tangent Function

When considering the ***inverse tangent function*** (or the ***arctan function***) we first restrict the domain of the tangent function to $-\frac{\pi}{2} < \theta < \frac{\pi}{2}$.

The inverse tangent function takes as its input the slope of the terminal ray (a tangent value) and outputs a single angle measure that produces that tangent value.

We write this function as $\theta = \arctan(m)$ or $\theta = \tan^{-1}(m)$.

---

e. Complete the function machine diagrams by identifying the input and output quantities for each function and the domain and range of each function.

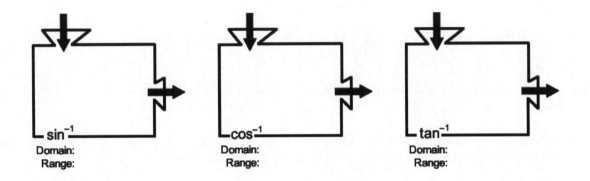

sin⁻¹
Domain:
Range:

cos⁻¹
Domain:
Range:

tan⁻¹
Domain:
Range:

*2. a. Demonstrate on the following circles how we can estimate the value of $\sin^{-1}(0.25)$.

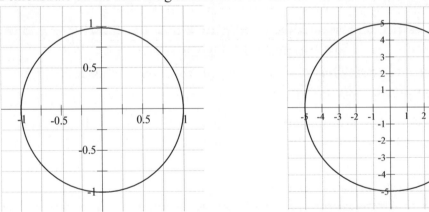

b. Use your calculator to evaluate $\sin^{-1}(0.25)$. Explain why the answer your calculator gives makes sense.

c. Your calculator only returns one possible value for $\sin^{-1}(0.25)$. What other angle measures have a sine value of 0.25? Give at least two.

---

*3. a. Demonstrate on the following circles how we can estimate the value of $\cos^{-1}(-0.75)$.

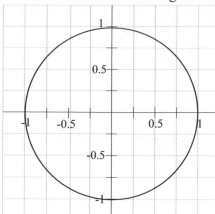 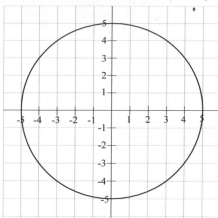

b. Use your calculator to evaluate $\cos^{-1}(-0.75)$. Explain why the answer your calculator gives makes sense.

c. Your calculator only returns one possible value for $\cos^{-1}(-0.75)$. What other angle measures have a cosine value of -0.75? Give at least two.

*4. a. Demonstrate on the following circles how we can estimate the value of $\tan^{-1}(2)$.

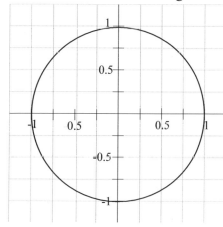 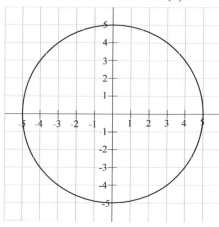

b. Use your calculator to evaluate $\tan^{-1}(2)$. Explain why the answer your calculator gives makes sense.

c. Your calculator only returns one possible value for $\tan^{-1}(2)$. What other angle measures have a tangent value of 2? Give at least two.

*5. Savanna is sitting in a bucket of a Ferris wheel at the 3 o'clock position. The Ferris wheel has a radius 62 feet long. The Ferris wheel begins moving counterclockwise.

   a. What is the measure of Savanna's angle of rotation when she is 42 feet above the Ferris wheel's horizontal diameter?

   b. What is the measure of Savanna's angle of rotation when she is –20 feet to the right of the Ferris wheel's vertical diameter?

6. Determine the angle measure θ (in radians) indicated on the diagram.

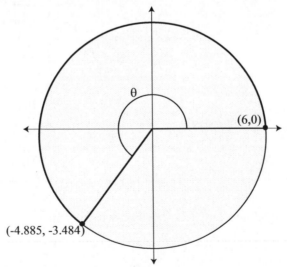

*7. A skier skied on a circular trail, starting at the position (2.5, 0) on the circle and ending at the position (–2.3775, 0.7725).

  a. Draw a picture of this situation.

  b. How many kilometers did the skier ski? Calculate this distance using both the inverse sine and inverse cosine functions.

8. A circle is centered at the origin and has a radius 4 meters long. An arc is drawn from (4, 0) counterclockwise to each indicated point. Find the length of each arc (in the same units as the coordinates).

  a.   (1.874, –3.534) meters   b.   (–0.837, –0.547) radii

*9. Determine *values* of θ , $0 \le \theta \le 2\pi$, that make each statement true. If there are no such values of θ , say "no solution" and explain how you know.

  a. $3\cos(\theta) - 1$ b. $\frac{1}{4}\cos(\theta) = -0.2$ c. $5\cos(\theta) = -5$

d. $2\sin(\theta) = 1$

e. $\sin(\theta) = .8$

f. $\frac{1}{3}\sin(\theta) = 0.5$

g. $\tan(\theta) = 1$

h. $3\tan(2\theta) = -8$

i. $2\tan(\theta) = -\frac{1}{3}$

*10. Use the given diagram to complete the following. Assume all angles are measured in radians.

a. What is the value of θ ?

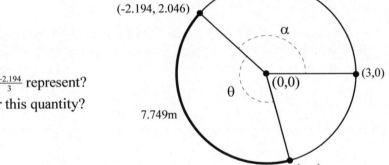

(-2.194, 2.046)

(3,0)

(0,0)

7.749m

θ

α

(x, y)

b. What quantity does the value $\frac{-2.194}{3}$ represent? What is the unit of measure for this quantity?

c. What quantity does the value $\frac{2.046}{3}$ represent? What is the unit of measure for this quantity?

d. What is the value of α ?

e. Determine the coordinates for the point labeled (x, y).

I. ANGLE MEASURE (TEXT: S1)

1. a. Describe the meaning of an angle measuring 1 radian. Construct an angle and illustrate the angle's measure on your diagram.
    b. Describe the meaning of an angle measuring θ radians. Construct an angle and illustrate the angle's measure on your diagram.

2. Explain, in terms of arc length and circumference, what it means for an angle to have a measure of:
    a. 45 degrees
    b. 191.4 degrees

3. The "grad" is a unit of angle measure that is sometimes used in France. One grad is the measure of an angle that cuts off an arc that is $\frac{1}{400}$ th of a circle's circumference. An angle that rotates a circle's circumference is 400 grads. (*Recall that one degree is an angle with vertex at the origin that cuts off an arc that is $\frac{1}{360}$ th of a circle's circumference and every circle's circumference is 360 degrees.*)
    a. Using measures of circumference and arc length, explain how to make a protractor that measures an angle's openness in grads.
    b. If the protractor has a 4.3-inch radius, determine the arc length in inches of:
        i. 1 grad   ii. 10 grads   iii. 200 grads   iv. 400 grads
    c. An angle measuring 50 grads subtends what percent of a circle's circumference? Explain.
    d. How many degrees are equivalent to 50 grads? 100 grads? 10.2 grads?
    e. Define a function that converts a number of grads to a number of degrees. Explain how you determined this function.

4. Name and define your own unit of angle measurement (e.g., describe how many of these units mark off any circle's circumference, making it at least 10 units). Describe how to create a protractor to measure angles using the unit you defined.

5. Is there a benefit to our choice of units for measuring angles? For example, is there a benefit to using 360 degrees or 400 grads to rotate a circle compared to some unit that cuts off an arc that is $\frac{1}{13}$ th or $\frac{1}{70}$ th of a circle's circumference?

6. Explain in terms of arc length and radius what it means for an angle to have a measure of:
    a. 1 radian   b. 2.1 radians   c. 3 radians   d. π radians   e. $\frac{1}{2}\pi$ radians

7. a. Convert the following angle measures from degrees to radians. Explain your method and why it works.
        i. 37°   ii. 310°   iii. 715°   iv. 90°
    b. Convert the following angle measures from radians to degrees. Explain your method and why it works.
        i. 3.7 radians   ii. 0.5π radians   iii. 6.28 radians   iv. 2π radians   v. 5.5 radians
    c. Define a function f that accepts an angle measure in degrees as input and returns a unique angle measure in radians as output. Define your variables.
    d. Define a function g that accepts an angle measure in radians as input and returns a unique angle measure in degrees as output. Define your variables.
    e. Describe the process of composing f and g. Evaluate $g(f(x))$.
    f. If the openness of an angle varies from 35° to 112°, the openness of the angle changes by how many radians? Draw a circle and illustrate the change in angle measure.

8. The *barc* is a unit of angle measure. One barc is the measure of an angle that cuts off an arc that is $\frac{1}{240}$ th of a circle's circumference. An angle that rotates a circle's circumference is 240 barcs.

   a. Using measures of circumference and arc length, explain how to make a protractor that measure an angle's openness in barcs.

   b. If a protractor for measuring angles in barcs has a 7.1-inch radius, explain how a circle can be used to create this protractor. What is the arc length in inches on this protractor of 1 barc? 50 barcs? 100 barcs?

   c. How many radians are equivalent to 200 barcs? 521.2 barcs?

   d. Define a function that converts a number of degrees to a number of barcs. Define a function that converts a number of barcs to a number of radians.

   e. Using the two functions above, determine a function that converts a number of degrees to a number of radians.

II. ANGLE MEASURE IN CONTEXT (TEXT: S2)

9. a. Determine the percentage of a circle's circumference cut off by an angle that has a measure of 22.3 degrees plus 301.1 grads (recall that an angle subtending the entire circumference of a circle measures 360 degrees or 400 grads).

   b. The measure of the angle is: i) how many degrees? ii) how many radians?

10. Three angles have their vertex at the center of a circle with a circumference of 21 inches. Their measures are given below. Determine the linear measure, in inches, of the arc cut off by each angle. Make a sketch to illustrate each measure.

    a. 192 degrees   b. 4.7 radians   c. $\frac{\pi}{6}$ radians

11. a. Determine the measure of an angle (in radians) that cuts off an arc length of 5.1 inches in a circle with a 5.1-inch radius.

    b. Determine the measure of an angle (in radians) that cuts off an arc length of 10.2 inches in a circle with a 5.1-inch radius.

    c. Determine the measure of an angle (in radians) that cuts off an arc length of 21.3 inches in a circle with a 5.1-inch radius.

12. Answer the following questions.

    a. Determine the value of *s*, the arc length (measured in inches) cut off by the angle with a measure of 2.1 radians.

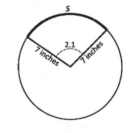

    b. Determine angle measure θ in radians. Then determine the equivalent angle measure in degrees.

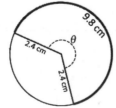

    c. Determine the radius (measured in feet) of the circle with the given measures.

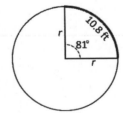

13. April is riding on a circular Ferris wheel that has a 51-foot radius. After boarding the Ferris wheel she traveled a distance of 32.2 feet along an arc before the Ferris wheel stopped for the next rider.
    a. Make a drawing of the situation and illustrate the relevant quantities.
    b. The angle that April swept out along the arc had a measure of: i) how many degrees? ii) how many radians?

14. Consider an object moving on a circular path with 4.2-meter radius.
    a. How many radians are swept out when the object travels 19 meters on the circular path? Make a drawing of the situation and illustrate the quantities.
    b. How many meters does the object travel when it sweeps out 47 degrees on the circular path? Make a drawing of the situation and illustrate the quantities.
    c. If the object sweeps out $\frac{\pi}{7}$ radians on the circular path, how many meters has it traveled?
    d. Suppose the distance the object traveled on the circular path varied from 3.2 meters to 8.3 meters. How many radians did the object sweep out over this distance? Make a drawing of the situation and illustrate the quantities.

15. Using the given diagram, define a formula that relates the measures r, θ, and s. Consider r and s to be linear measures with the same units, like *number of inches*, and θ to be radians. Explain how you determined this relationship.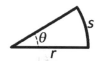

16. A common formula to calculate an angle's measure is $\theta = \frac{s}{r}$, where θ represents the measure of an angle's openness in radians, s represents the linear measure of arc length, and r represents the radius in the same units as s. Make a drawing of the situation and illustrate the quantities. Explain why the number of radius lengths in an arc length $\frac{s}{r}$ cut off by the angle results in an angle measure in radians.

III. MODELING CIRCULAR MOTION USING THE SINE AND COSINE FUNCTION (TEXT: S3)

For Exercises #17-20, imagine a bug sitting on the end of a blasé of a fan as the blade revolves in a counter-clockwise direction. The bug is exactly 2.9 feet from the center of the fan and is at the 3 o'clock position as the blade begins to turn.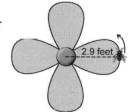

17. What quantities might you track to represent the location of a bug as it sleeps on the tip of a rotating fan blade? List as many quantities as necessary.

18. If the bug travels 6.8 feet around the circle from the 3 o'clock position, what angle measure (in radians) has been swept out?

19. a. If the bug swept out 1.23 radius lengths on an arc as the fan blade rotated, how many feet did the bug travel along the arc?
    b. If the bug swept out 1.23 radius lengths on an arc while sitting on a tip of a fan blade with a longer radius, would the bug travel the same number of feet as in part (a)? Explain.
    c. If a bug travels $2r$ feet (where r represents the length of the fan blade of various sized fans), how many radius lengths has the bug swept out along the arc of each of these fans?

20. a. Sketch a graph of the bug's vertical distance above the horizontal diameter (measured in feet) in terms of the measure of the angle swept out by the bug's fan blade (in radians). Label the horizontal axes with "Number of radians swept out by the bug's fan blade" and begin by plotting points. *Keep in mind that the vertical distance is positive when the bug is above the horizontal diameter line and negative when the bug is below it.*

   b. i. Use your graph in part (a) to sketch a second graph that also illustrates the bug's vertical distance above the horizontal diameter in terms of the measure of the angle swept out by the bug's fan blade (in radians), but for this graph, have the output be expressed in a number of radius lengths instead of feet.

   ii. What is the maximum value that the bug is away from the horizontal diameter in radius lengths? In feet?

   iii. What is the relationship between 1 radius and 2.9 feet for this situation?

   iv. Why are the shapes of the graphs the same?

   c. Explain how the bug's vertical distance above the horizontal diameter changes as the bug rotates from: i) the 3 o'clock position to the 12 o'clock position; ii) the 12 o'clock position to the 9 o'clock position; iii) the 9 o'clock position to the 6 o'clock position; iv) the 6 o'clock position to the 3 o'clock position.

   d. How does the graph you constructed in part (a) change if the radius of the fan is 3.9 feet instead of 2.9 feet?

   e. Given that the relationship in part (b) is modeled by the function $f(\theta) = \sin(\theta)$, define a function g to model the relationship in part (a).

   f. For successive equal change of angle measure from the 6 o'clock position to the 3 o'clock position, how do the corresponding changes in the vertical distance change? Identify these changes on a diagram of the situation and on your graph in part (b).

For Exercises #21-22, Ana is sitting in the bucket of a Ferris wheel. She is exactly 46.7 feet from the center and is at the 3 o'clock position as the Ferris wheel starts turning.

21. a. If Ana travels 2.58 radius lengths on a circular arc as the Ferris wheel rotates, how many feet did she travel? Define a function to express the number of feet Ana travels on this Ferris wheel (with a 46.7-foot radius) as a function of the angle measure θ.

   b. Ana's sister is sitting on a Ferris wheel that has longer arms than Ana's does. The sister travels 2.58 radius lengths in her ride. Did the sister travel the same number of feet as Ana? If not, who traveled farther? Explain.

22. a. Sketch a graph of Ana's horizontal distance to the right of the vertical diameter (measured in feet) in terms of the measure of the angle swept out by Ana (in radians). Label the horizontal axes with "Number of radians swept out by Ana" and begin by plotting points. (*The horizontal distance is positive when Ana is to the right of the vertical diameter line, and negative when Ana is to the left of the vertical diameter line.*)

   b. i. Sketch a second graph that also illustrates how Ana's horizontal distance to the right of the vertical diameter in terms of the measure of the angle swept out by Ana in radians, but for this graph have the output be express in a number of radius lengths instead of feet.

   ii. What is the maximum distance that Ana is away from the vertical diameter in radius lengths and in feet?

   iii. What is the relationship between 1 radius and 46.7 feet for this situation?

   iv. Why are the shapes of the graphs the same?

   c. Which of the two graphs will change if the radius of the Ferris wheel is changed? Explain.

Exercise continues on the next page.

d.   Explain how Ana's horizontal distance to the right of the vertical diameter changes as Ana rotates from: i) the 3 o'clock position to the 12 o'clock position; ii) the 12 o'clock position to the 9 o'clock position; iii) the 9 o'clock position to the 6 o'clock position; iv) the 6 o'clock position to the 3 o'clock position.

e.   Given that the graph in part (b) is represented by the function $f(\theta) = \sin(\theta)$, define a function g to represent the graph in part (a).

f.   For successive equal changes of angle measure from the 9 o'clock position to the 6 o'clock position, how do the corresponding changes in the horizontal distance change? Identify these changes on both a diagram of the situation and on your graph in part (b).

23. a.   Fill in the blanks of the table for the function $f(\theta) = \sin(\theta)$.

As the input θ varies from...	the function $g(\theta)$ (increases, decreases, or remains constant)	from a value of ...	to a value of...	resulting in a change of ... in f's output
0 to π/6				
π/6 to π/3				
π/3 to π/2				
π/2 to π				
π to 3π/2				
3π/2 to 5π/3				
5π/3 to 11π/6				
11π/6 to 2π				

b.   What patterns do you notice about how the value of $f(\theta)$ varies as θ varies from 0 to 2π?

c.   How does the change in $f(\theta)$ vary as θ varies by small, equal amounts on the interval from $\frac{\pi}{2}$ to $\frac{3\pi}{2}$?

24. a.   Fill in the blanks for the table for the function $g(\theta) = \cos(\theta)$.

As the input θ varies from...	the function $f(\theta)$ (increases, decreases, or remains constant)	from a value of ...	to a value of...	resulting in a change of ... in g's output
0 to π/6				
π/6 to π/3				
π/3 to π/2				
π/2 to π				
π to 3π/2				
3π/2 to 5π/3				
5π/3 to 11π/6				
11π/6 to 2π				

b.   What patterns do you notice about how the value of $g(\theta)$ varies as θ varies from 0 to 2π?

c.   How does the change in $g(\theta)$ vary as θ varies by small, equal amounts on the interval from π to 2π?

d.   What conclusions can you make about the inflection points on g's graph, and the concavity of g's graph, as θ varies from π radians to 2π radians?

IV. USING THE SINE AND COSINE FUNCTIONS TO TRACK CIRCULAR MOTION (TEXT: S4)

25. Determine the point (x, y) indicated on the circle.

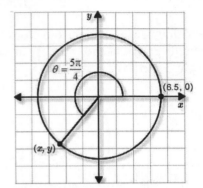

26. Determine the point (x, y) indicated on the circle.

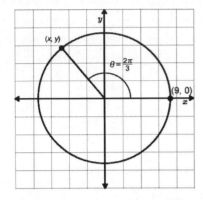

27. This figure illustrates the path of a toy racecar that begins at $(4, 0)$ and travels 19 meters counter-clockwise on a circular path with a 4-meter radius. The racecar stops at the point (x, y).

    Represent the values of x and y using trigonometric functions and then determine the decimal approximations.

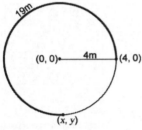

28. This figure illustrates the path of a toy racecar that begins at $(8, 0)$ and travels 38 meters counter-clockwise on a circular path with an 8-meter radius. The racecar stops at the point (x, y).
    a. Determine the angle measure (in radians) that subtends the 38-meter arc.
    b. Represent the values of x and y using trigonometric functions and then determine the decimal approximations.
    c. Define a formula that represents the horizontal component x (in radians) in terms of the number of meters d the racecar has traveled along the track.
    d. Define a formula that represents the vertical component y (in radians) in terms of the number of meters d the racecar has traveled along the track.
    e. Define a formula that represents the horizontal component x (in meters) in terms 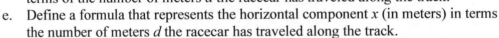 of the number of meters d the racecar has traveled along the track.
    f. Define a formula that represents the vertical component y (in meters) in terms of the number of meters d the racecar has traveled along the track.

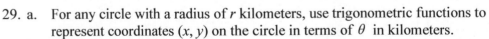

29. a. For any circle with a radius of r kilometers, use trigonometric functions to represent coordinates (x, y) on the circle in terms of θ in kilometers.
    b. Represent the coordinates (x, y) in *radius lengths* (instead of kilometers)

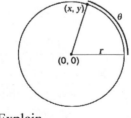

30. Answer the following questions:
    a. Determine $\sin\left(\frac{\pi}{6}\right)$ and $\sin\left(\frac{5\pi}{6}\right)$. What do you notice about the two values? Explain.

    b. Determine $\sin\left(\frac{7\pi}{6}\right)$ and $\sin\left(\frac{11\pi}{6}\right)$ without a calculator. Illustrate your answers on the unit circle.

    c. Given that $\cos\left(\frac{\pi}{6}\right) = \frac{\sqrt{3}}{2}$, determine $\cos\left(\frac{5\pi}{6}\right)$ without a calculator. Illustrate your answer on a unit circle.

---

d. Given that $\cos\left(\frac{\pi}{3}\right) = \frac{1}{2}$, determine $\cos\left(\frac{4\pi}{3}\right)$ without a calculator. Illustrate your answer on a unit circle.

31. For parts (a) through (d), construct a unit circle and use it to determine your answers.

a. Determine all possible values of θ that make the equation $\sin(\theta) = \frac{\sqrt{2}}{2}$ true. What is the value of $\cos(\theta)$ for each of these values of θ ?

b. Determine all possible values of θ that make the equation $\cos(\theta) = \frac{-1}{2}$ true. What is the value of $\sin(\theta)$ for each of these values of θ ?

c. Determine all possible values of θ that make the equation $\cos(\theta) = 0$ true. What is the value of $\sin(\theta)$ for each of these values of θ ?

d. Determine all possible values of θ that make the equation $\sin(\theta) = -1$ true. What is the value of $\cos(\theta)$ for each of these values of θ ?

32. For parts (a) through (d), assume we want measures of θ such that $0 \le \theta < 2\pi$.

a. Given $\sin(\theta) = 0.707$ and that $\cos(\theta)$ is negative, determine the value of θ and the value of $\cos(\theta)$. Then draw a diagram to represent your answer on a unit circle.

b. Given $\cos(\theta) = -0.99$ and that $\sin(\theta)$ is positive, determine the value of θ and the value of $\sin(\theta)$. Then draw a diagram to represent your answer on a unit circle.

c. Given $\cos(\theta) = -0.99$ and that $\sin(\theta)$ is negative, determine the value of θ and the value of $\sin(\theta)$. Then draw a diagram to represent your answer on a unit circle.

d. Given $\sin(\theta) = -0.91$ and that $\cos(\theta)$ is negative, determine the value of θ and the value of $\cos(\theta)$. Then draw a diagram to represent your answer on a unit circle.

V. USING THE SINE AND COSINE FUNCTIONS IN APPLIED SETTINGS (TEXT: S5)

33. Jim travels 2 radius lengths per second as he rotates on a Ferris wheel (starting in the 3 o'clock position). Let t represent the number of seconds elapsed since the Ferris wheel started rotating.

a. An angle measuring θ radians subtends the arc he travels. How does θ change as t varies? *We suggest first drawing a diagram of this situation.*

b. Define a function v that represents Jim's *vertical distance above* the horizontal diameter of the Ferris wheel in terms of t.

c. Define a function h that represents Jim's *horizontal distance to the right* of the vertical diameter of the Ferris wheel in terms of t.

d. Over what time interval does Jim complete one full revolution?

34. Repeat Exercise #33 if Jim instead travels 0.5 radius lengths per second.

35. Repeat Exercise #33 if Jim instead travels $\frac{1}{3}$ radius lengths per second.

36. Matthew determined that it takes 75 seconds for him to complete one revolution on a Ferris wheel.

a. Determine the speed that Matthew travels (in radius lengths per second).

b. Determine a function m that represents Matthew's *vertical distance above* the horizontal diameter of the Ferris wheel in terms of the number of seconds elapsed since he started to rotate counter-clockwise from the 3 o'clock position. *We suggest first drawing a diagram of the situation.*

c. Sketch a graph of the function m and mark the interval of input values on which the function values complete one full cycle.

---

37. Bill extends his arm straight out holding a 2.2-foot string with a ball attached to the end. He twirls the ball around in a circle with his hand at the center, so that the plane in which it is twirling is perpendicular to the ground.
    - Assume that the ball twirls counter-clockwise starting at the 3 o'clock position.
    - Let n represent the number of seconds since the ball starting rotating from the 3 o'clock position.
    - Let θ represent the measure of the angle (in radians) subtending the arc the ball has traveled.
    a. Make a diagram of the situation and illustrate the quantities. Label the ball's starting position.
    b. Define a function f that represents the *vertical distance* of the ball *above* Bill's hand (measured in feet) in terms of θ.
    c. Suppose Bill swings the ball at a constant rate such that it makes 3 revolutions in 5 seconds. At what rate is the ball traveling in feet per second? In radius lengths per second?
    d. Write a formula that represents the distance the ball traveled d (measured in feet) in terms of n.
    e. Write a formula that represents θ in terms of n.
    f. Sketch a graph of the relationship in part (d). On your graph, illustrate the interval of time (the values of n) for which the ball completes one revolution.
    g. Define a function g that represents the *vertical distance* of the ball *above* Bill's hand (measured in feet) in terms of n.

38. Caren extends her arm straight out holding a 1.3-meter string with a ball attached to the end. She twirls the ball around in a circle with her hand at the center, so that the plane in which it is twirling is perpendicular to the ground.
    - Assume that the ball twirls counter-clockwise starting at the 3 o'clock position.
    - Let n represent the number of seconds since the ball starting rotating from the 3 o'clock position.
    - Let θ represent the measure of the angle (in radians) subtending the arc the ball has traveled.
    a. Make a diagram of the situation and illustrate the quantities. Label the ball's starting position.
    b. Define a function f that represents the *horizontal distance* of the ball *to the right of* Caren's hand (measured in meters) in terms of θ.
    c. Suppose Caren swings the ball at a constant rate such that it makes 1 revolutions in 0.7 seconds. At what rate is the ball traveling in meters per second? In radius lengths per second?
    d. Write a formula that represents the distance the ball traveled d (measured in meters) in terms of n.
    e. Write a formula that represents θ in terms of n.
    f. Define a function g that represents the *horizontal distance* of the ball *to the right of* Caren's hand (measured in meters) in terms of n.
    g. Sketch a graph of the relationship in part (f). On your graph, illustrate the interval of time (the values of n) for which the ball completes one revolution.
    h. Suppose that the ball travels 0.25 radius lengths per second. Define a function j that represents the ball's horizontal distance (measured in meters) to the right of Caren's hand in terms of the number of seconds since the ball began to move. Over what time interval did the ball complete one revolution?

VI. TRANSFORMATIONS OF THE SINE AND COSINE FUNCTIONS (TEXT: S6)

39. Given that f is defined by $f(\theta) = \sin(\theta)$, and g is defined by $g(\theta) = \sin(2\theta)$,
    a. By how much must θ vary so that 2θ varies by 2π radians?
    b. Determine g's period.
    c. How does the graph of g compare with the graph of f?

40. Given that f is defined by $f(\theta) = \sin(\theta)$, and h is defined by $h(\theta) = \sin\left(\frac{\theta}{4}\right)$,

    a. By how much must θ vary so that $\frac{\theta}{4}$ varies by 2π radians?

    b. Determine h's period.

    c. How does the graph of h compare with the graph of f ?

41. Given that f is defined by $f(\theta) = \cos(\theta)$, and j is defined by $j(\theta) = \sin\left(\frac{2\theta}{5}\right)$,

    a. By how much must θ vary so that $\frac{2\theta}{5}$ varies by 2π radians?

    b. Determine j's period.

    c. How does the graph of j compare with the graph of f ?

42. a. Sketch a graph of $h(x) = \sin(5x)$, where x is in radians, on the interval $-2\pi < x < 2\pi$.

    b. What is the maximum value for $h(x)$? For what value(s) of x does $h(x)$ assume its maximum value? (*Give exact answers in terms of* π.)

    c. Identify the root(s) and vertical intercept of h. (*Express your answers in term of* π *if appropriate.*)

    d. Identify the period of h on the graph.

    e. Describe how the output values of h vary as the input values vary from $\frac{\pi}{2}$ to π radians.

    f. What is the amplitude of h?

43. a. Sketch a graph of $g(x) = 2.5\sin\left(\frac{x}{2}\right)$, where x is in radians, on the interval $-2\pi < x < 8\pi$.

    b. What is the maximum value of $g(x)$? For what value(s) of x does $g(x)$ assume its maximum value? (*Give exact answers in terms of* π.)

    c. Identify the root(s) and vertical intercept of g. (*Express your answers in terms of* π *if appropriate.*)

    d. Identify the period of g on the graph.

    e. Describe how the output values of g vary as the input values vary from $\frac{\pi}{2}$ to π radians.

    f. What is the amplitude of g?

44. Let $f(\theta) = \sin(3\theta)$, where θ is in radians and is any real number. What is the period of f ? How does the graph of f compare to the graph of $p(\theta) = \sin(\theta)$? Justify your answer.

45. Let $g(\theta) = \cos\left(\frac{4\theta}{9}\right)$, where θ is in radians and is any real number. What is the period of g? How does the graph of g compare to the graph of $q(\theta) = \cos(\theta)$? Justify your answer.

46. Let $h(\theta) = \cos(8.2\theta)$, where θ is in radians and is any real number. How does the graph of h compare to the graph of $q(\theta) = \cos(\theta)$? Justify your answer.

47. Consider the three functions $f(x) = \cos(x)$, $g(x) = \cos(0.1x)$ and $h(x) = \cos(3\pi x)$. All inputs are in radians.

    a. Over what interval of x will x vary by 2π?

    b. Over what interval of x will $0.1x$ vary by 2π?

    c. Over what interval of x will $3\pi x$ vary by 2π?

    d. Sketch graphs of each of the three functions on the same set of axes. Illustrate the period of each function on its graph.

48. Compare the periods of $f(x) = \sin(\frac{1}{2}x)$, $g(x) = \sin(x)$, and $h(x) = \sin(2x)$. Sketch graphs of the three functions on the same set of axes and label their periods.

49. Let $f(x) = 2.1\cos(x)$ and $g(x) = \cos(x)$ with x measured in radians. Compare the functions' amplitudes, then graph both functions on the same axes and label their amplitudes.

50. Repeat Exercise #49 if $f(x) = 0.21\cos(x)$ and $g(x) = \cos(x)$.

51. a. Draw the graph of $f(x) = 0.8\sin(x)$, with x measured in radians, on the interval $-3\pi \le x < 3\pi$.
    b. What is the maximum value of $f(x)$? For what value(s) of x does $f(x)$ assume its maximum value?
    c. Identify the root(s) and vertical intercept of f.
    d. Identify the period of f on the graph.
    e. Describe how the output values of f vary as the input values vary from $\frac{\pi}{2}$ to π radians.
    f. What is the amplitude of f?

52. Repeat Exercise #51 for $f(x) = 5\sin(x)$ on the interval $-2\pi \le x < 3\pi$.

53. Repeat Exercise #51 for $f(x) = 5\cos(x)$ on the interval $0 \le x < 2\pi$.

54. Let $f(x) = \sin(x) + 1.7$ and $g(x) = \sin(x)$. Graph f and g on the same axes and compare their behavior.

55. Let $f(x) = 2\sin(x) + 1.6$ and $g(x) = \sin(x)$. For any x, how are $f(x)$ and $g(x)$ related?

56. Graphs of periodic functions are provided below. Illustrate the period of each function on its graph (approximations are okay), then determine a possible formula that defines the function.

a.
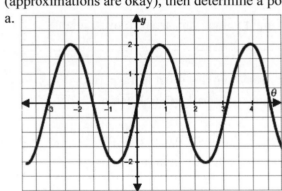

b.
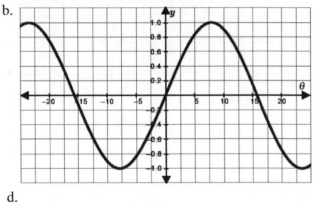

c.
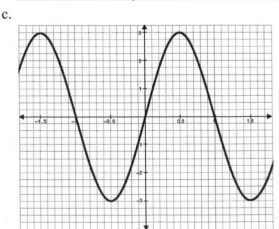

d.
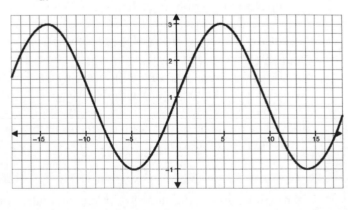

57. A diagram showing a water wheel (used for power in the 18th century) is given.

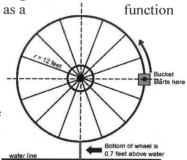

a. Sketch a graph of the *height of the bucket above the water* (in feet) as a function of the angle of rotation (measured in radians) from the 3 o'clock position. *Define your variables and label your axes.*

b. Define a function f that represents the bucket's height d above the water (in feet) in terms of the angle of rotation (measured in radians) from the 3 o'clock position.

c. Determine a function g that defines the bucket's height d above the water (in feet) in terms of the *number of feet* the bucket has rotated counter-clockwise from the 3 o'clock position.

58. The given graph models the average daily temperature (measured in degrees Fahrenheit) in Athens, GA as a function of the number of days passed since June 15th.

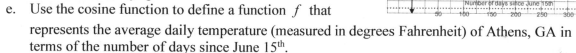

a. Choose a point on the graph and explain what it represents in the context.

b. What is the maximum average daily temperature in Athens, GA? On what day does the maximum average temperature occur?

c. What is the minimum average temperature in Athens, GA? On what day does the minimum average temperature occur?

d. How does the average daily temperature change over the time period from June 15th to December 15th? Explain your answer by discussing amounts of change in the average daily temperature for successive equal changes in the number of days passed since June 15th.

e. Use the cosine function to define a function f that represents the average daily temperature (measured in degrees Fahrenheit) of Athens, GA in terms of the number of days since June 15th.

f. Explain what each constant used in the function define in part (e) represents in this context.

59. An object travels on a circular path perpendicular to the ground, starting at the 3 o'clock position and moving counterclockwise. The angle measure subtending the object's path traveled (in radians) as a function of time (measured in seconds) is given by the function $f(t) = t^2$. Thus, the object's vertical distance (measured in radii) above the center of the circular path as a function of time (measured in seconds) is given by the function $g(t) = \sin(t^2)$.

a. Sketch a graph of the function f for $t \geq 0$. For successive equal changes of 2π in the *output* of f, what is happening to the corresponding changes in input?

b. Sketch a graph of the function g for $t \geq 0$. Explain how your findings in part (a) are reflected in the sketch of the graphs of f and g.

VII. SHIFTS/TRANSFORMATIONS OF PERIODIC FUNCTIONS (TEXT: S7)

60. Explain why the graph of g defined by $g(x) = (x-4)^3$ is shifted 4 units to the right of the graph of f defined by $f(x) = x^3$.

61. Determine how the graph of $g(\theta) = \sin\left(\theta - \frac{\pi}{2}\right)$ compares to the graph of $f(\theta) = \sin(\theta)$ by first setting up a table of values for both f and g, then plot their graphs on the same axes. Provide an explanation to support your answer.

62. Explain why the graph of $g(\theta) = \cos(\theta + \pi)$ is shifted π units left of the graph of $f(\theta) = \cos(\theta)$.

63. Shawna sits on a Ferris wheel at the 9 o'clock position and the Ferris wheel begins to rotate counter-clockwise. The Ferris wheel has a 25-foot radius and its center is 30 feet off the ground.

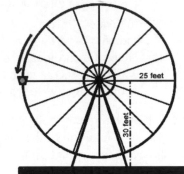

25 feet

30 feet

   a. After the Ferris wheel rotates π radians (half of one full rotation), what is her height above the ground?

   b. Your classmate says that the expression $25\sin(\pi) + 30$ should calculate Shawna's height above the ground (in feet) after the Ferris wheel rotates π radians. Do you agree? Explain.

   c. After the Ferris wheel rotates through each of the following rotation angles, what is Shawna's height above the ground? *Write an expression that represents her height using the sine function and then provide a decimal approximation of her height above the ground (in feet). We recommend drawing a diagram of each situation first.*

      i. $\frac{\pi}{2}$ radians   ii. π radians   iii. $\frac{\pi}{3}$ radians   iv. $\frac{3\pi}{4}$ radians

   d. Define the formula for function f that represents Shawna's height above the ground (in feet) in terms of the measure of the rotation angle since she boarded (θ, in radians).

   e. Graph f.

   f. If the Ferris wheel is rotating at 0.25 radians per second, define the formula for a function that represents Shawna's height above the ground (in feet) in terms of the number of seconds since the Ferris wheel started to rotate.

   g. If it takes the Ferris wheel 60 seconds to make one full rotation, define the formula for a function that represents Shawna's height above the ground (in feet) in terms of the number of seconds since the Ferris wheel started to rotate.

64. Construct the graph of h given that $h(\theta) = \sin(4\theta) + 3$. How does the graph of h compare to the graph of $f(\theta) = \sin(\theta)$?

65. Construct the graph of g given that $g(\theta) = \frac{2}{3}\cos\left(\frac{\theta}{2} - \pi\right)$. How does the graph of g compare to the graph of $f(\theta) = \cos(\theta)$?

66. Explain why the output of f defined by $f(x) = \cos\left(x - \frac{\pi}{2}\right)$ is equal to the output of g defined by $g(x) = \sin(x)$ for all values of x. (*Hint: Use the unit circle to justify that* $\cos\left(x - \frac{\pi}{2}\right) = \sin(x)$ *for all values of x.*)

67. Explain how the function f defined by $f(x) = \sin\left(x + \frac{\pi}{2}\right)$ is related to the function g defined by $g(x) = \cos(x)$. Sketch graphs of f and g on the same set of axes and identify the period and amplitude of both functions on their graphs.

68. Determine the amplitude and period of the following trigonometric functions without graphing them. Sketch a graph of each function after determining its period and amplitude then label the period and amplitude. (*Hint: You will need to consider the effect of some functions' values in addition to their amplitude and period*). All inputs are in radians.

   a. $f(t) = 3.2\sin(2t)$ b. $g(x) = 0.7\cos(2x + \pi)$ c. $h(x) = \sin(0.4x) + 1.21$

   d. $s(t) = 3\sin\left(\frac{2\pi}{21}t\right) + 2$ e. $q(c) = 2\cos\left(\frac{\pi}{9}c\right) - 7$ f. $r(t) = 0.1\sin(t - 5) + \pi$

69. Given the graphs of f and g, define a formula that defines g in terms of f.

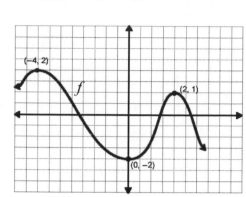
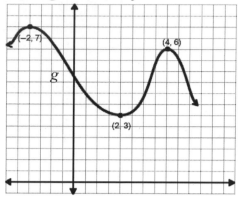

70. Graphs of periodic functions are provided below. Illustrate the period and amplitude of each function on its graph (you may approximate the period), then determine a formula that defines the function.

a.
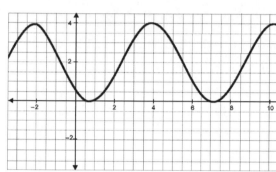

b.
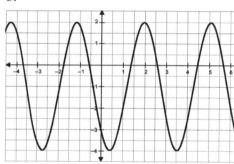

c.
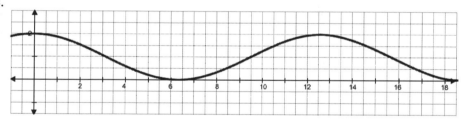

VIII.   THE TANGENT FUNCTION (TEXT: S8)

71. Evaluate each of the following then interpret the meaning of your answer.

a. $\tan(0)$ b. $\tan\left(\frac{11\pi}{6}\right)$ c. $\tan\left(\frac{\pi}{2}\right)$ d. $3\tan\left(\frac{1}{2}\right)$ e. $\dfrac{\sin\left(\frac{\pi}{3}\right)}{\cos\left(\frac{\pi}{3}\right)}$ f. $\dfrac{\sin\left(\frac{5\pi}{4}\right)}{\cos\left(\frac{5\pi}{4}\right)}$

72. Draw an angle with a measure of θ that meets the set of requirements given.

a. $\tan(\theta)$ is positive and $\cos(\theta)$ is negative

b. $\sin(\theta)$ is positive and $\tan(\theta)$ is negative

c. $\cos(\theta)$ is positive and $\tan(\theta)$ is negative

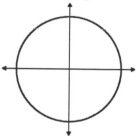
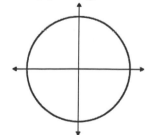
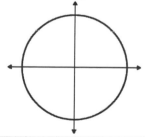

73. What values of θ make the equation $\tan(\theta) = 0$ true when $0 \le \theta < 4\pi$?

74. Let $f(\theta) = \tan(\theta)$. What is the input variable for the inverse relation of the function f and what does it represent? What is the output variable for the inverse relation of the function f and what does it represent?

75. The graph of $f(\theta) = \tan(\theta)$ is given. Use your understanding of function translations to make a rough sketch of a graph for each of the following functions.

   a. $g(\theta) = \tan(2\theta)$

   b. $h(\theta) = \tan\left(\frac{\theta}{3}\right)$

   c. $j(\theta) = 2\tan(\theta)$

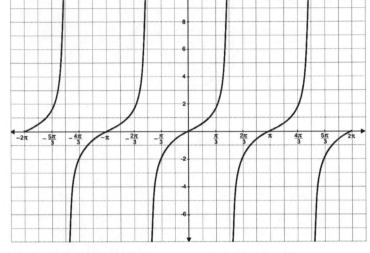

Since we interpret the tangent function as returning the slope of the terminal ray for an angle measuring θ radians, we can also define the identity $\tan(\theta) = \frac{y}{x}$ with (x, y) representing the intersection point of the terminal ray and the circle. Use this information for Exercises #76-77.

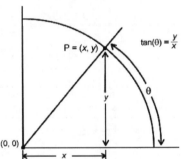

76. Explain why the identity $\tan(\theta) = \frac{y}{x}$ must be true.

77. Let θ vary from 0 to $\frac{\pi}{2}$.

   a. How does x vary?
   b. How does y vary?
   c. How does the ratio $\frac{y}{x}$ vary?
   d. Explain how the variation of x and y on the interval from $\theta = 0$ to $\theta = 1$ impacts the variation of $\tan(\theta) = \frac{y}{x}$ on this same interval of θ.

IX. NEGATIVE ANGLE MEASURE, CO-TERMINAL ANGLES, AND PERIODICITY

There is no printed homework for this investigation.

X. INVERSE TRIGONOMETRIC FUNCTIONS (TEXT: S9)

78. Given that $\sin\left(\frac{\pi}{6}\right) = \frac{1}{2}$, determine the value of $\arcsin\left(\frac{1}{2}\right)$. Justify your answer.

79. If $\cos\left(\frac{\pi}{3}\right) = \frac{1}{2}$, determine the value of $\cos^{-1}\left(\frac{1}{2}\right)$. Justify your answer.

80. If $\tan\left(\frac{\pi}{4}\right) = 1$, determine the value of $\tan^{-1}(1)$. Justify your answer.

81. If $\tan\left(\frac{2\pi}{3}\right) = -\sqrt{3}$, determine the value of $\tan^{-1}\left(-\sqrt{3}\right)$. Justify your answer.

82. If $\cos\left(\frac{5\pi}{4}\right) = -\frac{\sqrt{2}}{2}$, determine the value of $\cos^{-1}\left(-\frac{\sqrt{2}}{2}\right)$. Justify your answer.

83. Explain why $\sin^{-1}\left(\sin(\theta)\right) = \theta$ when $-\frac{\pi}{2} \le \theta \le \frac{\pi}{2}$.

84. For what values of θ is the statement $\tan^{-1}\left(\tan(\theta)\right) = \theta$ true? Explain your reasoning.

85. Explain why $\cos\left(\cos^{-1}\left(\frac{3\theta}{2}\right)\right) = \frac{3\theta}{2}$ when $-\frac{2}{3} \le \theta \le \frac{2}{3}$.

86. Demonstrate on a circle how you can estimate the value of $\cos^{-1}(-0.8)$.

87. Evaluate $\sin^{-1}(0.7)$. What does your answer represent?

88. Evaluate the following expressions in radians.
    a. $\arccos(0.5)$ b. $3\arcsin(-1)$ c. $5\arccos(0)$

89. Evaluate each of the following for $x = 0.5$.
    a. $\sin\left(x^{-1}\right)$ b. $\left(\sin(x)\right)^{-1}$ c. $\sin^{-1}(x)$
    d. Explain the meaning conveyed by the notation in each of parts (a) through (c).

90. a.   Solve the following equations for values of θ, given that $-\frac{\pi}{2} < \theta < \frac{\pi}{2}$
       i. $\tan(\theta) = 1$ ii. $\tan(\theta) = -1$ iii. $3\tan(2\theta) = -8$ iv. $2\tan(\theta) = -\frac{1}{3}$
    b. In part (a), why was it necessary to restrict the domain to $-\frac{\pi}{2} < \theta < \frac{\pi}{2}$ in order to provide a single answer in parts (i) through (iv)?

91. Solve the following equations for θ if possible. If not, say why not.
    a. $3\sin(\theta) = 2$ b. $\cos(\theta) = -0.3$ c. $\frac{1}{2}\sin(\theta) = -0.2$

92. Solve each equation for θ, given that $0 \le \theta < 2\pi$. Be sure to show your work. Check your answer by graphing and labeling your solutions on your graphs.
    a. $\sin(4\theta) = 0.5$ b. $2\cos(\theta) = -\frac{1}{3}$ c. $\sin(\theta) = 0$ d. $8\cos(2\theta) = -6$

93. Determine the measure θ of the angle (in radians) indicated on the circle.

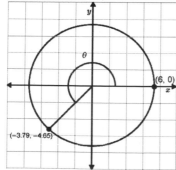

94. Determine the measure θ of the angle (in radians) indicated on the circle.

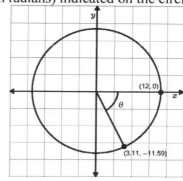

95. Given the following points on a circle with a 4-meter radius centered at the origin, determine the corresponding arc length (in the same units as the coordinates of each point) between (4, 0) and the given point (rotated counter-clockwise).
    a. (–0.713, 3.936) meters
    b. (–0.924, 0.382) radii
    c. (–3.282, –2.286) meters
    d. (–0.924, –0.382) radii

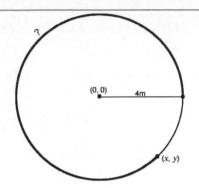

96. Given the following circle and undetermined angle measures α and θ radians, answer the following questions.
    a. What is the value of θ ?
    b. What quantity does the value $\frac{1.3028}{5}$ represent? What is the unit of measure for this quantity?
    c. What quantity does the value $\frac{4.8273}{5}$ measure? What is the unit of measure for this quantity?
    d. What is the value of α ?

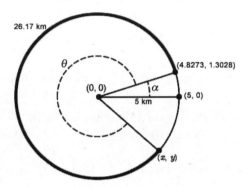

97. An arctic village maintains a circular cross-country ski trail that has a 2.5-km radius. A skier started skiing from position (–1.76777, –1.76777), measured in kilometers, and skied counter-clockwise for 3.927 kilometers, where he paused for a brief rest. Determine the ordered pair (in both kilometers and radii) on the coordinate axes that identifies the location where the skier rested.

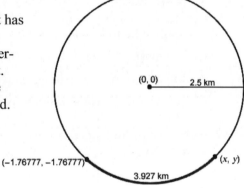

98. A skier started skiing from position (2.4136, 0.6513), measured in kilometers, and skied counter-clockwise for 13.09 kilometers where he paused for a brief rest. Determine the ordered pair (in both kilometers and radii) on the coordinate axes that identifies the location where the skier rested.

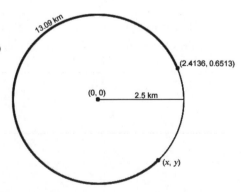

99. Porter is sitting in a bucket of a Ferris wheel at the 3 o'clock position. The Ferris wheel has a 45-foot radius. The Ferris wheel begins moving counterclockwise.

   a. Define a function to represent Porter's vertical distance (in feet) above the horizontal diameter of the Ferris wheel in terms of his angle of rotation (in radians) as he rotates counter clockwise from the 3 o'clock position

   b. What is the angle of rotation when Porter is 27 feet above the Ferris wheel's center? (*Set up an equation and show how to use the arcsine or* $\sin^{-1}$ *function to determine one of the angles.*)

   c. Define a function to represent Porter's horizontal distance to the right of the Ferris wheel's center in terms of the angle of rotation (in radians).

   d. What angle(s) has Porter rotated when he is -5 feet to the right of the Ferris wheel's center? (*Set up an equation and show how to use* $\cos^{-1}$ *to solve for one of the angles.*)

100. A biologist tracked the deer population in a rural area of Wisconsin and found that the deer population in this area was cyclic. He used his data to find a function to approximate the population over a year-long period. The function g defined by $g(t) = 1125 - 875\cos\left(\frac{\pi}{6}t\right)$, $0 \le t < 12$, represents the number of deer $g(t)$ in terms of the number of months t since November 1, 2009. Use the arccosine function to determine the month(s) in which the deer population was at least 600.

101. Wind turbines, or windmills, generate energy with less pollution than traditional power plants.

   a. Choose values for the height of the turbine's tower (measured in meters) and for the turbine's radius (in meters - do not choose "1"). Label these on the diagram.

   b. Sketch a graph of the distance of the fan blade's tip (measured in meters) above the horizontal diameter of the windmill as a function of measure of the angle (in radians) swept out by the fan blade from a 3 o'clock starting position.

   c. Define a function f that represents the relationship you graphed in (b). Define variables to represent the values the quantities assume (include units).

   d. Sketch a graph of the distance of the fan blade's tip (measured in meters) *above the ground* as a function of the measure of the angle (in radians) swept out by the fan blade.

   e. Define a function g that represents the relationship you graphed in (d). Define variables to represent the values the quantities assume (include units).

   f. Explain how the function's definition in (e) reflects the fact that you are measuring distances *above the ground*.

102. A wind turbine blade rotates at a rate of 3 radians every 15 seconds.

   a. Define a function h that relates the measure of the angle (in radians) swept out by the fan blade as a function of time elapsed since the fan blade started rotating from the 3 o'clock position. Sketch the graph of h.

   b. On your graph in (a), identify the interval of time needed for the fan blade to complete one revolution.

   c. Sketch a graph of the distance of the fan blade's tip (in meters) above the horizontal diameter of the windmill as a function of the number of seconds that have elapsed since the fan started rotating from the 3 o'clock position.

   d. On your graph in (c), identify the interval of time needed for the fan blade to complete one revolution.

   e. Define a function f that will generate the graph you sketched in (c). Define variables to represent the values the quantities assume (include units).

   f. Explain how the function in (e) conveys the rate at which the fan blade rotates.

103. Monica and Louie go to the fair and decide to take a ride on the Ferris wheel. The Ferris wheel operator tells them that if they can keep track of how far they have traveled while in their seat, they will win a prize. The catch is that they have to be able to tell him how far they have traveled any time he asks for it. The only information he gives them is that the Ferris wheel's radius is 35 feet. After agreeing to his challenge, Monica notes there are 12 total carts equally spaced around the Ferris wheel. Louie and Monica are also the first group to board at the bottom of the Ferris wheel.
    a. Make a drawing of the situation and illustrate the quantities. In the diagram, label the location of Louie and Monica when boarding the Ferris wheel as well as the location of Louie and Monica when the last cart is filled.
    b. When the cart after their cart is being filled, what is the length of the arc (measured in feet) already traveled by Louie and Monica? Measured in radius lengths? Label these measurements on your diagram above, explicitly identifying the arc length being measured.
    c. After the operator has filled up four total carts, including Louie and Monica's, he asks them for the distance they have traveled. What is the length of the arc (measured in feet) traveled by Louie and Monica? Measured in radius lengths? Label these measurements on your diagram, explicitly identifying the arc length being measured.
    d. The Ferris wheel starts again and the operator continues to fill carts. Finally, as the last cart is being filled, the operator asks Louie and Monica how far they have traveled. What is the length of the arc (measured in feet) traveled by Louie and Monica? Measured in radius lengths? Label these measurements on your diagram above, explicitly identifying the arc length being measured.

104. Use the same context as Exercise #103.
    a. What was Louie and Monica's vertical distance *above the center of the Ferris wheel* (measured in feet) when they first boarded the Ferris wheel? Measured in radius lengths? (*Consider their vertical distance to be negative when below the center of the Ferris wheel and positive when above the center of the Ferris wheel.*)
    b. What is Louie and Monica's vertical distance *above the center of the Ferris wheel* (measured in feet) when the cart *after* Louie and Monica's cart is being loaded? Measured in radius lengths?
    c. Define a function f that relates Louie and Monica's vertical distance above the center of the Ferris wheel (measured in radius lengths) as a function of the angle of rotation (in radians).
    d. Define a function g that relates Louie and Monica's vertical distance above the center of the Ferris wheel (measured in feet) as a function of the angle of rotation (measured in radians).
    e. If the bottom of the Ferris wheel is 11 feet above the ground, define a function h that relates Louie and Monica's distance above the ground (measured in feet) as a function of the angle of rotation (measured in radians).
    f. Sketch a graph of the function h defined in part (e). Label two points on the graph and explain the meaning of these points in this context.
    g. Suppose the radius of the Ferris wheel were doubled. Would this change the function from part (e)? If so, re-write the function for this new Ferris wheel. Explain your reasoning.

105. Carlos is the last person to board a Ferris wheel that has a 52-foot radius. The counterclockwise arc swept out by Carlos is measured from where he boarded *at the bottom of the Ferris wheel*. Since Carlos is the last one to board, the Ferris wheel starts up and doesn't stop again until the ride is over.

   a. Draw a diagram that show's Carlos's location and his angle of rotation at an arbitrary moment in time.

   b. Define a function f that relates Carlos's *vertical* distance above the center of the Ferris wheel (measured in feet) as a function of his angle of rotation (in radians). Over what interval of input will Carlos complete one revolution?

   c. Define a function g that relates Carlos's *horizontal* distance to the right of the center of the Ferris wheel (measured in feet) as a function of his angle of rotation (measured in radians). Over what interval of input will Carlos complete one revolution?

   d. Suppose Carlos's cart is traveling 0.3 radius lengths per second on a circular path as the wheel rotates. Define a function h that relates Carlos's *horizontal* distance to the right of the center of the Ferris wheel (measured in feet) as a function of the time since beginning the ride (measured in seconds). How long will it take for Carlos to complete one revolution?

   e. Suppose that Carlos's cart traveled at 0.4 radius lengths per second. Define a function j that relates Carlos's *horizontal* distance to the right of the center of the Ferris wheel (measured in feet) as a function of the time since beginning the ride (measured in seconds). How long will it take for Carlos to complete one revolution?

   f. Suppose Carlos's cart is traveling 0.2 radius lengths per second. Define a function k that relates Carlos's *horizontal* distance to the right of the center of the Ferris wheel (measured in feet) as a function of the time since beginning the ride (measured in seconds). How long will it take for Carlos to complete one revolution?

   g. Sketch a graph of each function defined in parts (d) through (g). Label your axes. Identify the interval of input over which Carlos completes one revolution.

106. Lucia boards a Ferris wheel with a 42.5-foot radius. When Lucia boards, the Ferris wheel is at the bottom of its rotation. The angle swept out as Lucia rotates is measured from the 6 o'clock position.

   a. Make a drawing of the situation and illustrate the quantities.

   b. Graph Lucia's *vertical distance above* the ground (measured in feet) as a function of her angle of rotation (in radians). (*Define your variables and label your axes.*)

   c. Define a function f that relates Lucia's vertical distance above the ground (measured in feet) as a function of her angle of rotation (measured in radians) from the bottom of the Ferris wheel. (*Define variables to represent the values the quantities assume.*)

   d. Suppose that the Ferris wheel rotates so that each cart moves at a constant rate of 7.5 feet per second and makes no stops. Express the distance traveled by Lucia (in feet) as a function of the number of seconds since the Ferris wheel began to turn.

   e. Define a function g that relates Lucia's *vertical distance above* the ground (measured in feet) as a function of the number of seconds elapsed since the Ferris wheel began to turn. How many seconds will it take for Lucia to complete one revolution?

   f. Sketch a graph of the function you defined in part (e). On your graph, identify the interval of input values on which the function values complete one full cycle (i.e., the interval of t for which Lucia completes one revolution).

107. A buoy sitting in the ocean bobs up and down such that it moves vertically 4.2 feet between its high and low points every 9 seconds. When you peered out the window of your cabin the buoy was at its highest point and dropped to its lowest point before returning back to its highest point, completing the full cycle in 9 seconds.

    a. What quantities in this situation are changing? Define variables to represent the values the quantities assume (include units). Create a graph to represent the covariation of these two quantities. (*The graph should illustrate how the values of the two quantities change together.*) Be sure to label your axes.

    b. Define a function f that relates the vertical position of the buoy (measured in feet) in terms of the number of seconds that have elapsed since you began watching. Be sure to explicitly identify the input and output quantities of the situation.

    c. Describe what each variable and constant value in your function represents.

    d. Sketch a graph of your function from part (b) and describe the co-variation of the input and output quantities over a time period of 5 seconds to 6.1 seconds since you began watching.

    e. Use the function you graphed to determine the times when the buoy reaches the halfway position between its high and low points.

108. The London Eye is a large Ferris wheel that is a famous London landmark. The given function models a person's height above the ground (in feet) as a function of the number of minutes he/she has been on the Eye.

$$f(t) = -221\cos\left(\tfrac{\pi}{15}t\right) + 221$$

    a. What is the amplitude of f and what does this value represent in this context?

    b. What is the period of f and what does this value represent in the context of the problem? Explain how you determined the period.

    c. Define a function g that relates the height of a person above the ground (measured in feet) as a function of the distance traveled (measured in feet) using the cosine function.

    d. Alter the function f to reflect the situation in which the London Eye rotates twice as quickly.

    e. Alter the function f to reflect the situation in which the radius of the London Eye is doubled.

    f. Sketch graphs of the given function and the functions you defined in parts (d) and (e) on the same set of axes.

This investigation contains review and practice with important skills and procedures you may need in this module and future modules. Your instructor may assign this investigation as an introduction to the module or may ask you to complete select exercises "just in time" to help you when needed. Alternatively, you can complete these exercises on your own to help review important skills.

Rationalizing Denominators
Use this section prior to the module or with/after Investigation 1.

Mathematicians like to have standard ways of representing values of expressions because it makes communication easier and less prone to misunderstandings. For example, they tend to write variables after constants in an expression such as writing $2x$ instead of $x2$ to represent the product of 2 and x. It isn't <u>wrong</u> to write $x2$, but you might think it looks odd. This is because you are used to using the conventional notation.

When fractions contain an expression involving radicals in the denominator, their denominators are irrational numbers. In these cases mathematicians decided on the convention of *rationalizing* the denominator (making the denominator a rational number).

Consider the number $\frac{3}{\sqrt{5}}$. Multiplying by "1" does not change its value but can change its form. **See the work to the right.**

So $\frac{3}{\sqrt{5}}$ and $\frac{3\sqrt{5}}{5}$ are equivalent (they both represent a value of approximately 1.34164).

$\frac{3}{\sqrt{5}}$ has an irrational denominator. The denominator of $\frac{3\sqrt{5}}{5}$ has been *rationalized*.

$$\frac{3}{\sqrt{5}} = \frac{3}{\sqrt{5}} \cdot 1$$
$$= \frac{3}{\sqrt{5}} \cdot \frac{\sqrt{5}}{\sqrt{5}}$$
$$= \frac{3\sqrt{5}}{\sqrt{5} \cdot \sqrt{5}}$$
$$= \frac{3\sqrt{5}}{(\sqrt{5})^2}$$
$$= \frac{3\sqrt{5}}{5}$$

In Exercises #1-12, write an equivalent number with a rationalized denominator. *Reduce any fractions if possible. Feel free to check your work by finding and comparing the decimal approximations.*

1. $\dfrac{7}{\sqrt{10}}$

2. $\dfrac{9}{\sqrt{5}}$

3. $\dfrac{1}{\sqrt{11}}$

4. $\dfrac{3}{\sqrt{3}}$

5. $\dfrac{2}{\sqrt{10}}$

6. $\dfrac{4}{\sqrt{12}}$

7. $\dfrac{15}{\sqrt{5}}$

8. $\dfrac{20}{\sqrt{8}}$

9. $\dfrac{15}{8\sqrt{3}}$

10. $\dfrac{40}{7\sqrt{5}}$

11. $\dfrac{\sqrt{3}}{\sqrt{7}}$

12. $\dfrac{2\sqrt{10}}{3\sqrt{5}}$

One benefit of rationalizing denominators is that it makes it easier to create common denominators. For example, it is much easier to represent the sum $\frac{3\sqrt{3}}{\sqrt{28}} + \frac{5\sqrt{2}}{4\sqrt{5}}$ as a single fraction by finding common denominators after rationalizing them and writing the equivalent sum $\frac{3\sqrt{21}}{14} + \frac{\sqrt{10}}{4}$.

Special Right Triangles and Exact Values of Sine, Cosine
Use this section prior to the module or with/after Investigation 1.

Equilateral triangles have three congruent angles (so each angle must measure $\frac{180°}{3} = 60°$ or $\frac{\pi}{3}$ radians) and three congruent sides. Thus, every equilateral triangle looks similar to the figure on the right (where r represents any possible positive measurement for the side length in whatever units you choose).

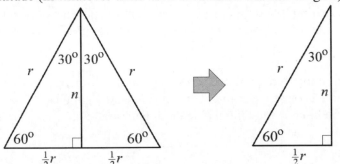

If we split this triangle in half, then we get the following. Let n represent the length of the altitude (in whatever units used to measure the side lengths).

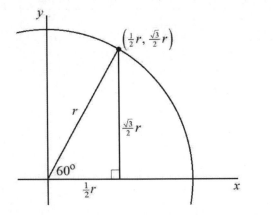

$$\left(\tfrac{1}{2}r\right)^2 + n^2 = r^2$$
$$\tfrac{1}{4}r^2 + n^2 = r^2$$
$$n^2 = r^2 - \tfrac{1}{4}r^2$$
$$n^2 = \tfrac{3}{4}r^2$$
$$n = \sqrt{\tfrac{3}{4}r^2}$$
$$n = \sqrt{\tfrac{3}{4}} \cdot \sqrt{r^2}$$
$$n = \tfrac{\sqrt{3}}{\sqrt{4}} \cdot r$$
$$n = \tfrac{\sqrt{3}}{2}r$$

Then we can use the Pythagorean theorem to determine the value of n in terms of r as shown. *Note that r and n must both be positive, which influences our calculations. For example, normally we would say that $\sqrt{r^2} = |r|$. However, since we know $r > 0$, we can say $\sqrt{r^2} = r$.*

Why is this important? Let's redraw the right triangle in two different orientations on a coordinate plane.

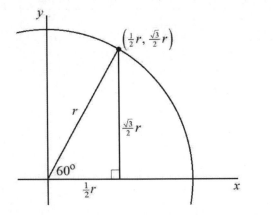
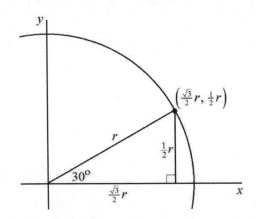

When the angle measures $60°$ (or $\frac{\pi}{3}$ radians), the point where the terminal ray intersects the circle is $\frac{1}{2}$ radii (or $\frac{1}{2}r$ units) to the right of the vertical axis and $\frac{\sqrt{3}}{2}$ radii (or $\frac{\sqrt{3}}{2}r$ units) above the horizontal axis.

When the angle measures $30°$ (or $\frac{\pi}{6}$ radians), the point where the terminal ray intersects the circle is $\frac{\sqrt{3}}{2}$ radii (or $\frac{\sqrt{3}}{2}r$ units) to the right of the vertical axis and $\frac{1}{2}$ radii (or $\frac{1}{2}r$ units) above the horizontal axis.

13. What does this tell us about the values of each of the following expressions?
    a. $\cos(30°)$

    b. $\sin(30°)$

    c. $\tan(30°)$

    d. $\cos(60°)$

    e. $\sin(60°)$

    f. $\tan(60°)$

14. What does this tell us about the values of each of the following expressions?
    a. $\cos\left(\frac{\pi}{6}\right)$

    b. $\cos\left(\frac{\pi}{3}\right)$

    c. $\sin\left(\frac{\pi}{6}\right)$

    d. $\sin\left(\frac{\pi}{3}\right)$

    e. $\tan\left(\frac{\pi}{6}\right)$

    f. $\tan\left(\frac{\pi}{3}\right)$

15. An angle's initial ray points in the 3 o'clock position and its terminal ray is rotated 30° (or $\frac{\pi}{6}$ radians) counterclockwise. The angle's vertex is at (0, 0), and a circle is also centered at (0, 0).
    a. Draw a picture of this scenario.

    b. Where does the terminal ray intersect the circle if...
       i. the circle's radius is 11 inches?

       ii. the circle's radius is 3.2 cm?

       iii. the circle's radius is r meters?

16. Repeat Exercise #15 if the angle measures 60° (or $\frac{\pi}{3}$ radians) instead.

A right isosceles triangle has one right angle and two congruent angles (that each measure $\frac{90°}{2} = 45°$ or $\frac{\pi}{4}$ radians) with two congruent side lengths. Let r represent the hypotenuse length and n represent the leg lengths (measured in the same units).

Let's determine the value of n in terms of r using the Pythagorean theorem. *Note that r and n must both be positive, which influences our calculations. For example, normally we would say that $\sqrt{r^2} = |r|$.*

However, since we know $r > 0$, we can say $\sqrt{r^2} = r$.

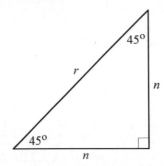

$$n^2 + n^2 = r^2$$
$$2n^2 = r^2$$
$$n^2 = \frac{r^2}{2}$$
$$n = \sqrt{\frac{r^2}{2}}$$
$$n = \frac{\sqrt{r^2}}{\sqrt{2}}$$
$$n = \frac{r}{\sqrt{2}}$$

By convention, mathematicians like to write this value in a different form by rationalizing the denominator.

$$n = \frac{r}{\sqrt{2}} \cdot \frac{\sqrt{2}}{\sqrt{2}}$$
$$= \frac{r \cdot \sqrt{2}}{\sqrt{2} \cdot \sqrt{2}}$$
$$= \frac{r\sqrt{2}}{2} \quad \text{or} \quad \frac{\sqrt{2}}{2} \cdot r$$

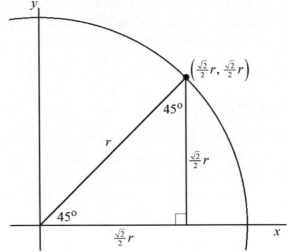

Let's redraw the right triangle on a coordinate plane (as shown to the right).

When the angle measures $45°$ (or $\frac{\pi}{4}$ radians), the point where the terminal ray intersects the circle is $\frac{\sqrt{2}}{2}$ radii (or $\frac{\sqrt{2}}{2}r$ units) to the right of the vertical axis and $\frac{\sqrt{2}}{2}$ radii (or $\frac{\sqrt{2}}{2}r$ units) above the horizontal axis.

17. What does this tell us about the values of each of the following expressions?
   a. $\cos(45°)$ b. $\sin(45°)$ c. $\tan(45°)$

   d. $\sin\left(\frac{\pi}{4}\right)$ e. $\tan\left(\frac{\pi}{4}\right)$ f. $\cos\left(\frac{\pi}{4}\right)$

18. An angle's initial ray points in the 3 o'clock position and its terminal ray is rotated 45° (or $\frac{\pi}{4}$ radians) counterclockwise. The angle's vertex is at (0, 0), and a circle is also centered at (0, 0).
    a. Draw a picture of this scenario.

    b. Where does the terminal ray intersect the circle if…
       i. the circle's radius is 15 cm?

       ii. the circle's radius is 4.5 inches?

       iii. the circle's radius is r feet?

Exact Trigonometric Values for Key Angle Measures in Quadrant I

The following are exact values for trigonometric functions at common key angle measures. *We provide the values based on degree measurements and radian measurements for the same angles.*

$$\cos(30°) = \frac{\sqrt{3}}{2} \qquad \sin(30°) = \frac{1}{2} \qquad \tan(30°) = \frac{\sqrt{3}}{3}$$

$$\cos\left(\frac{\pi}{6}\right) = \frac{\sqrt{3}}{2} \qquad \sin\left(\frac{\pi}{6}\right) = \frac{1}{2} \qquad \tan\left(\frac{\pi}{6}\right) = \frac{\sqrt{3}}{3}$$

$$\cos(45°) = \frac{\sqrt{2}}{2} \qquad \sin(45°) = \frac{\sqrt{2}}{2} \qquad \tan(45°) = 1$$

$$\cos\left(\frac{\pi}{4}\right) = \frac{\sqrt{2}}{2} \qquad \sin\left(\frac{\pi}{4}\right) = \frac{\sqrt{2}}{2} \qquad \tan\left(\frac{\pi}{4}\right) = 1$$

$$\cos(60°) = \frac{1}{2} \qquad \sin(60°) = \frac{\sqrt{3}}{2} \qquad \tan(60°) = \sqrt{3}$$

$$\cos\left(\frac{\pi}{3}\right) = \frac{1}{2} \qquad \sin\left(\frac{\pi}{3}\right) = \frac{\sqrt{3}}{2} \qquad \tan\left(\frac{\pi}{3}\right) = \sqrt{3}$$

Exact Values of Trigonometric Functions in Other Quadrants
Use this section prior to the module or with/after Investigation 3.

Once you know the exact values for trigonometric functions for key angles in Quadrant I, it's possible to know the exact values for trigonometric functions at other angles in Quadrants II-IV by determining the measure of the **reference angle** – the smallest angle formed by the terminal ray and the *x*-axis.

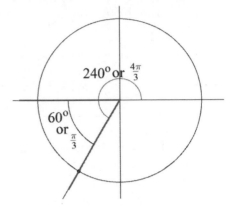

For example, consider the angle measuring 240° (or $\frac{4\pi}{3}$ radians). When we draw this angle, we can determine that the reference angle measures 60° (or $\frac{\pi}{3}$ radians). This is the measure of the smallest angle formed between the terminal ray and the *x*-axis.

The values of $\cos(240°)$, $\sin(240°)$, and $\tan(240°)$ will match the values of $\cos(60°)$, $\sin(60°)$, and $\tan(60°)$ **EXCEPT** perhaps for their signs. Since the original angle terminates in Quadrant III, we know that the cosine and sine values must both be negative and that the tangent value must be positive. Thus, we get the following.

$$\cos(240°) = -\tfrac{1}{2}, \quad \sin(240°) = -\tfrac{\sqrt{3}}{2}, \quad \cos(240°) = \sqrt{3}$$

or

$$\cos\left(\tfrac{4\pi}{3}\right) = -\tfrac{1}{2}, \quad \sin\left(\tfrac{4\pi}{3}\right) = -\tfrac{\sqrt{3}}{2}, \quad \cos\left(\tfrac{4\pi}{3}\right) = \sqrt{3}$$

In Exercises #19-30 you are given the measure of an angle θ with its initial ray pointing in the 3 o-clock direction and its terminal ray rotated counterclockwise. Do the following.
    a) Sketch a diagram representing the situation on another sheet of paper.
    b) Determine the measure of the reference angle.
    c) Determine the exact values of $\cos(\theta)$, $\sin(\theta)$, and $\tan(\theta)$.

19. $\theta = 315°$ 20. $\theta = 150°$ 21. $\theta = 120°$ 22. $\theta = 225°$

23. $\theta = 330°$ 24. $\theta = 210°$ 25. $\theta = \frac{7\pi}{6}$ 26. $\theta = \frac{2\pi}{3}$

27. $\theta = \frac{3\pi}{4}$ 28. $\theta = \frac{5\pi}{3}$ 29. $\theta = \frac{5\pi}{4}$ 30. $\theta = \frac{11\pi}{6}$

*1. Recall that the measure of an angle corresponds to an arc length subtended on a circle.

    a. By using the triangle's hypotenuse as a radius, draw a circle with its center at the angle's vertex. Without first determining the measure of θ, determine the value of $\sin(\theta)$.

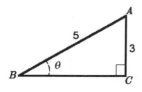

    b. By using the triangle's hypotenuse as a radius, draw a circle with its center at the angle's vertex. Without first determining the value of θ, determine the value of $\sin(\theta)$.

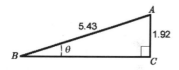

    c. By using the triangle's hypotenuse as a radius, draw a circle with its center at the angle's vertex. Without first determining the measure of θ, determine the length of $\overline{BC}$.

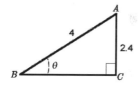

    d. Determine the values of $\cos(\theta)$ and $\sin(\theta)$. Explain your reasoning.

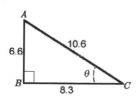

e. By using the triangle's hypotenuse as a radius,
   draw a circle with its center at the angle's
   vertex. Without first determining the value of θ,
   determine the value of $\tan(\theta)$.

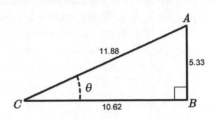

f. By using the triangle's hypotenuse as a radius,
   draw a circle with its center at the angle's
   vertex. Then determine the unknown angle
   measures.

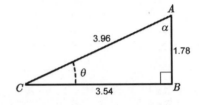

*2. Determine each of the following values.
   a. $\sin(\alpha)$ b. $\cos(\alpha)$

   c. $\tan(\alpha)$ d. $\sin(\beta)$

   e. $\cos(\beta)$ f. $\tan(\beta)$

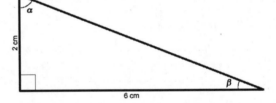

   g. Determine the angle measures α and β.

*3. a. Given that $\sin(\theta) = -\frac{3}{4}$ and that the angle measuring θ terminates in Quadrant III, determine
       $\cos(\theta)$ and $\tan(\theta)$.

   b. Given that $\tan(\alpha) = -\frac{2}{3}$ and that the angle measuring α terminates in Quadrant II, determine
       $\sin(\alpha)$ and $\cos(\alpha)$.

*1. A person standing 12 feet from a light pole notices that the light casts a shadow of his body that is 5 feet long. The individual is 5 feet, 6 inches tall
    a. Create a diagram and label the known and unknown quantity measures.

    b. Determine the height of the light pole and justify your solution.

    c. Determine an alternative method to find the height of the light pole. (*For example, if you used (inverse) trigonometric functions to solve part (b), use similar triangles to determine the unknown height. If you used similar triangles to solve part (b), use (inverse) trigonometric functions to determine the unknown height.*)

*2. A plane leaves a local air force base and travels due east. A radar station 45 miles south of the base tracks the plane and determines that the angle formed by the base, the radar station, and the plane is initially changing by 1.6 degrees per minute.
    a. Sketch a diagram of the situation and label the known and unknown quantities.

    b. Define a function that relates the plane's distance from the radar station d (measured in miles) as a function of the elapsed time t (measured in minutes) since leaving the base.

    c. Define a function that relates the plane's distance from the air force base c (measured in miles) as a function of the elapsed time t since leaving the base (measured in minutes).

d. Use your function in part (c) to complete the table.

Number of minutes t since leaving the base	The plane's distance c (measured in miles) from the base	$\dfrac{\Delta c}{\Delta t}$ (miles per minutes)
0		
0.5		
1		
1.5		
2		

e. Use your function in part (c) to complete the following table.

Number of minutes t since leaving the base	The plane's distance c (measured in miles) from the base	$\dfrac{\Delta c}{\Delta t}$ (miles per minutes)
42		
45.5		
49		
52.5		
56		

f. Describe how the plane's distance from the air force base c changes as a function of the elapsed time between 0 and 2 minutes and then again from 42 and 56 minutes since leaving the base. Are these values realistic? Why or why not?

g. Graph the relationship between the plane's distance from the air force base and the amount of time since leaving the base. What does the graph and the tables in parts (d) and (e) convey about the speed of the plane?

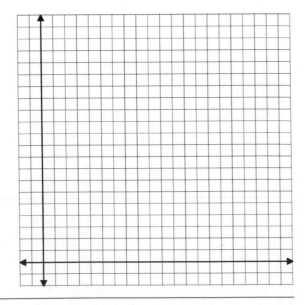

1. Use the given triangle to answer the following questions.
   a. Use the Pythagorean Theorem to express the relationship between x, y, and r.

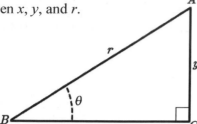

   b. Determine a formula relating $\cos(\theta)$ and x.

   c. Determine a formula relating $\sin(\theta)$ and y.

   d. Rewrite the relationship determined in part (a) using only $\cos(\theta)$, $\sin(\theta)$, and r.

2. Rewrite the following expressions in their most simplified form to generate an identity. *Make sure to state the values of x that create undefined expression values.*

   a. $\dfrac{\sin(x)}{\csc(x)} + \dfrac{\cos(x)}{\sec(x)}$

   b. $\dfrac{\tan(x)\cot(x)}{\csc(x)}$

   c. $\sin(x)\tan(x) + \cos(x)$

3. Algebraically prove the following identities. *In other words, choose the expression on one side of the equals sign and simplify this expression to equal the expression on the other side of the equals sign. Make sure to state the values of x that create undefined expression values.*

   a. $1 + \tan^2(x) = \sec^2(x)$

   b. $\frac{1 - \cot(x)}{\cos(x)} = \sec(x) - \csc(x)$

   c. $\left(\sec(x) - \tan(x)\right)^2 = \frac{1 - \sin(x)}{1 + \sin(x)}$

4. a. Use the trigonometric identity $\cos(x + y) = \cos(x)\cos(y) - \sin(x)\sin(y)$ to determine the exact value of $\cos\left(\frac{\pi}{12}\right)\cos\left(\frac{2\pi}{3}\right) - \sin\left(\frac{\pi}{12}\right)\sin\left(\frac{2\pi}{3}\right)$.

   b. Use the trigonometric identity $\cos(x - y) = \cos(x)\cos(y) + \sin(x)\sin(y)$ to determine the exact value of $\cos\left(\frac{13\pi}{16}\right)\cos\left(\frac{\pi}{16}\right) + \sin\left(\frac{13\pi}{16}\right)\sin\left(\frac{\pi}{16}\right)$.

*1. Surveyors were interested in determining the approximate size of a lake, so they set up their instruments in three different locations and measured the angles created. They were also able to directly measure the distance between Surveyor A and Surveyor B as shown.

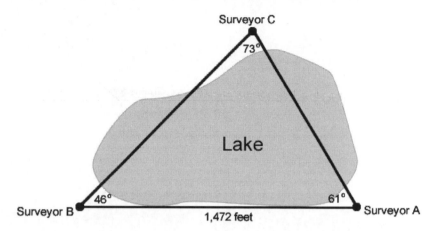

a. The triangle created is not a right triangle. However, we can apply the techniques of right triangle trigonometry by forming right triangles out of the given triangle. Draw a diagram below to show a few different ways that this is possible.

b. The lines you drew in part (a) to create right triangles are called *altitudes* (lines drawn from a vertex of a triangle perpendicular to the opposite side). Which altitude(s) will create a diagram that allows us to apply right triangle trigonometry techniques to solve for the lengths of the remaining sides of the triangle? Defend your choice.

c. Find the distance from Surveyor A to Surveyor C and from Surveyor B to Surveyor C.

*2. Two ships have each determined their distances from a lighthouse on the shore. An observer at the lighthouse determines the angle between the ships. Determine the distance between the ships.

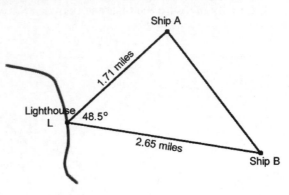

*3. Find the unknown side lengths and unknown angle measures in each triangle.

a.

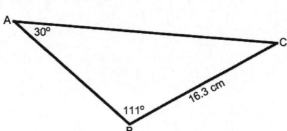

b.

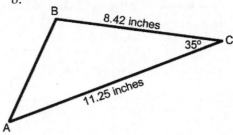

*4. Find the area of the following triangle.

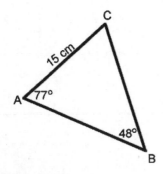

*1. A non-right triangle ABC has side lengths a, b, and c units with an altitude $\overline{BD}$ with a length of h units.

a. Write an expression that calculates the value of h if $m\angle A$ and c are known.

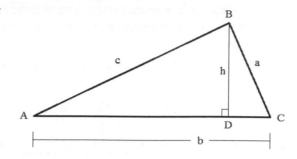

b. Write an expression that calculates the value of h if $m\angle C$ and a are known.

c. What do the results of parts (a) and (b) tell us?

2. Using the same triangle, we draw the altitude $\overline{CE}$ with a length of k units.

a. Write an expression that calculates the value of k if $m\angle A$ and b are known.

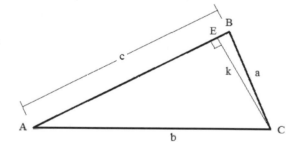

b. Write an expression that calculates the value of k if $m\angle B$ and a are known.

c. What do the results of parts (a) and (b) tell us?

Since the triangle in Exercises #1-2 was generic (no specific side lengths or angle measures were given), our conclusions will generalize to *any* triangle. This generalization produces the rule known as **The Law of Sines**.

The Law of Sines

For any non-right triangle *ABC*, we have that

$$\frac{a}{\sin(m\angle A)} = \frac{b}{\sin(m\angle B)} = \frac{c}{\sin(m\angle C)}.$$

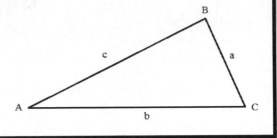

---

3. Use the Law of Sines to find the value of *x* in each of the following triangles.

   a.

   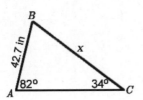

   b.

   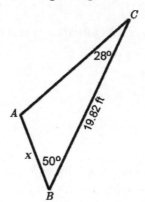

*4. Use the Law of Sines to find the values of *x* and *y* in each of the following triangles.

   a.

   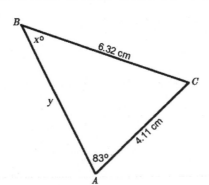

   b.

   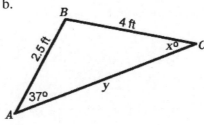

*5. In many cases the Law of Sines works perfectly well and returns the correct missing values in a non-right triangle. However, in some cases the Law of Sines returns *two* possible measurements. This is the case in the given triangle.

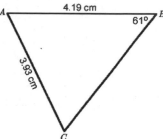

We begin the solution process, shown here.

$$\frac{4.19}{\sin(m\angle C)} = \frac{3.93}{\sin(61°)}$$

$$\frac{\sin(m\angle C)}{4.19} = \frac{\sin(61°)}{3.93}$$

$$\sin(m\angle C) = \frac{\sin(61°)}{3.93} \cdot 4.19$$

$$\sin(m\angle C) \approx 0.9325$$

$$m\angle C \approx \sin^{-1}(0.9325)$$

It's at this point that we get two possible answers for $m\angle C \approx \sin^{-1}(0.9325)$.

a. Why does $m\angle C \approx \sin^{-1}(0.9325)$ return two possible answers for $m\angle C$ in this triangle?

b. If we assume that the drawing is more or less to scale, which value of $m\angle C$ makes more sense? Use this value of $m\angle C$ to determine the remaining measurements of the triangle.

c. Determine the dimensions of the second possible triangle with the given measurements, then draw the triangle approximately to scale.

6. There are also two possible triangles with the given measurements shown below. Find the measurements of both triangles and draw each triangle approximately to scale.

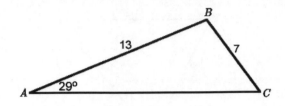

*1. The Law of Sines is an extremely useful tool for finding missing lengths or missing angle measures in non-right triangles. However, there are times when there isn't enough information to apply the Law of Sines. Examine the following non-right triangles and explain why the Law of Sines cannot be used to find missing side lengths or missing angle measures.

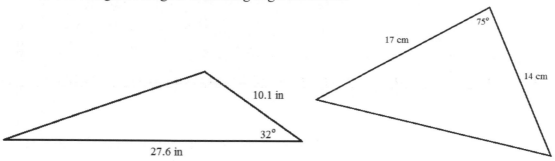

Let's take a situation like the triangles shown in Exercise #1 (*a triangle in which the measure of one angle is known and the length of the included sides are known*). For example, suppose we know $m\angle C$ and the lengths a and b in the following triangle and we are interested in finding the length c.

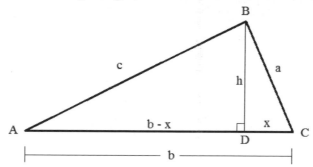

First, observe that the Pythagorean theorem assures that $h^2 + (b-x)^2 = c^2$ and $x^2 + h^2 = a^2$ (so $h^2 = a^2 - x^2$). Also, we have that $\cos(m\angle C) = \dfrac{x}{a}$, so $x = a \cdot \cos(m\angle C)$. Then

$$h^2 + (b-x)^2 = c^2$$
$$h^2 + (b-x)(b-x) = c^2$$
$$h^2 + b^2 - 2bx + x^2 = c^2$$
$$(a^2 - x^2) + b^2 - 2bx + x^2 = c^2 \qquad \bullet \text{ since } h^2 = a^2 - x^2$$
$$a^2 - \cancel{x^2} + b^2 - 2bx + \cancel{x^2} = c^2$$
$$a^2 + b^2 - 2bx = c^2$$
$$a^2 + b^2 - 2b(a \cdot \cos(m\angle C)) = c^2 \qquad \bullet \text{ since } x = a \cdot \cos(m\angle C)$$
$$a^2 + b^2 - 2ab \cdot \cos(m\angle C) = c^2$$

The Law of Cosines

For any non-right triangle ABC, we have that
$a^2 + b^2 - 2ab \cdot \cos(m\angle C) = c^2$.

Depending on the known information, equivalent
forms of this property include:

$$a^2 = b^2 + c^2 - 2bc \cdot \cos(m\angle A)$$
$$b^2 = a^2 + c^2 - 2ac \cdot \cos(m\angle B)$$
$$c^2 = a^2 + b^2 - 2ab \cdot \cos(m\angle C)$$

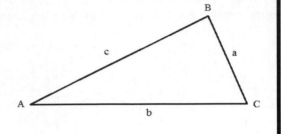

---

2. Use the Law of Cosines to find the value of x in each of the following triangles.

   *a.

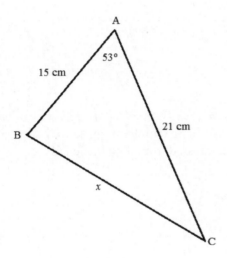

   b.

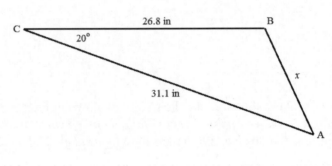

---

3. Use the Law of Cosines to find the value of *x* in each of the following triangles.

*a.

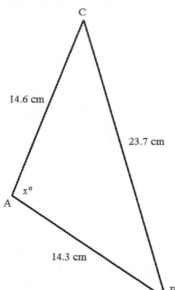

b.

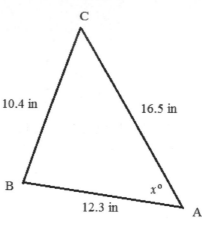

*4. The Law of Cosines and the Law of Sines can be applied to other shapes as well by first breaking them into triangles. The diagram below shows how this may be done.

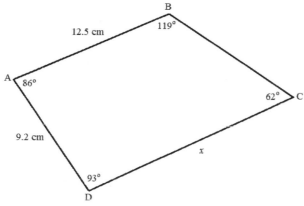 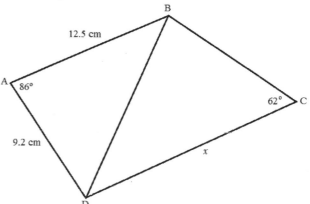

Use the given information to find the value of *x*.

5. Find the value of x in the given figure.

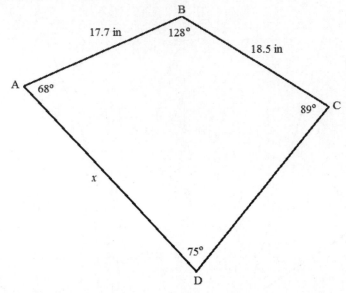

I. RIGHT TRIANGLE TRIGONOMETRY

1. Given a circle centered at the origin and a point (x, y) on the circle in the first quadrant, state the meaning of $\sin(\theta)$, $\cos(\theta)$, and $\tan(\theta)$ in terms of x and y; then in terms of o, h, and a.

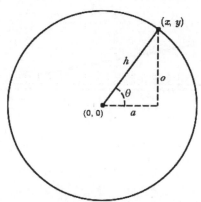

2. Determine the measure of the angles (in both degrees and radians) and the length of each side for each right triangle below.

a.

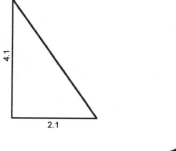

b.

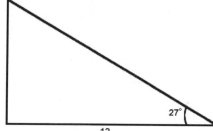

c.

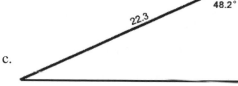

d.

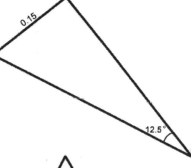

e.

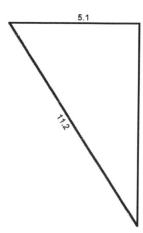

f.

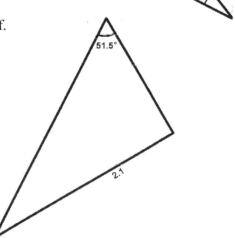

3. Use the diagram to complete the following.
   a. Find the length of hypotenuse $\overline{BC}$.
   b. Find the length of side $\overline{AB}$.
   c. Find $m\angle C$.

4. Use the diagram to complete the following.
   a. Find the length of the hypotenuse $\overline{RT}$.
   b. Find $m\angle R$.
   c. Find $m\angle T$.

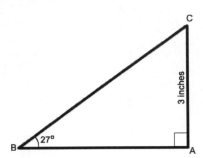

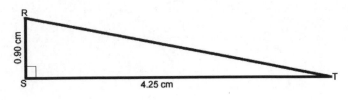

5. Determine the missing measurements. Assume θ is measured in radians.
   a.

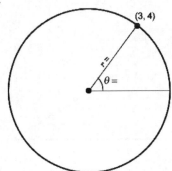

   b.

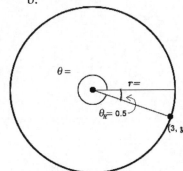

   c.

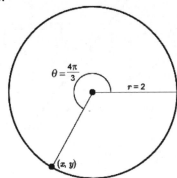

   d.

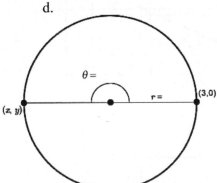

6. a. Given that $\cos(\theta) = \frac{1}{3}$ and that the angle measuring θ terminates in Quadrant I, determine $\sin(\theta)$ and $\tan(\theta)$.

   b. Given that $\tan(\alpha) = -\frac{2}{5}$ and that the angle measuring α terminates in quadrant IV, determine $\sin(\alpha)$ and $\cos(\alpha)$.

II. RIGHT TRIANGLE TRIGONOMETRY APPLICATIONS

7. A 10-foot ladder is leaning against a wall with the bottom of the ladder positioned 2 feet from the wall. A painter comes along and pulls the ladder away from the wall at a constant rate of 0.75 feet per second.
   a. Fill in the values of the following table.

t	x	y	θ *(degrees)*	$\dfrac{\Delta y}{\Delta t}$	$\dfrac{\Delta \theta}{\Delta t}$
0					
1					
2					
3					
4					
5					
6					
7					
8					
9					
10					
10.67					

   b. Describe how the distance of the bottom of the ladder from the wall x (in feet) changes with respect to the elapsed time t measured in seconds.
   c. Describe how the height of the top of the ladder from the ground y (in feet) changes with respect to the elapsed time t measured in seconds.
   d. Describe how the angle measure between the ladder and the ground θ (in degrees) changes with respect to the elapsed time t measured in seconds.
   e. Define a function f that relates the vertical distance, y, as a function of the horizontal distance, x.
   f. Define a function g that relates the angle measure θ as a function of the horizontal distance x.
   g. Define a function h that relates the horizontal distance, x, as a function of the elapsed time t.
   h. Define a function j that relates the angle measure, θ, as a function of the elapsed time t.

8. After Jack planted his magic beans, his neighbor Jill watched the beanstalk grow. When the top of the beanstalk was at her eye level (5 feet), Jill began tracking the growth of the beanstalk from a position 27 feet from the base of the beanstalk. After 125 seconds, she noted that the beanstalk reached the first cloud and estimated that her line of site was at an 85° angle with respect to the ground.
   a. Draw a diagram of the situation and label each known and unknown quantity measure.
   b. How tall was the beanstalk 125 seconds after Jill began tracking the growth of the beanstalk?
   c. How fast did the beanstalk grow, given that the beanstalk grew vertically at a constant rate of change of distance with respect to time?
   d. Define a function that relates the height of the beanstalk from the ground h as a function of the angle of Jill's line of sight with respect to the ground as she watched the top of the stalk grow.

9. A platform is to be built on a tall pole. To keep the platform from swaying in the wind, a total of eight guy-wires will be attached to the pole. The pole has a height of 166 feet. Four guy-wires will be attached to the pole at a height of 140 feet. These wires are 161 feet in length. The other four guy-wires will be attached to the pole at a height of 109 feet. These guy-wires are 131 feet in length.
   a. Draw a diagram of the situation and label each known and unknown quantity measure.
   b. At what distance from the base of the pole will the longer guy-wires be fastened to the ground?
   c. At what distance from the base of the pole will the shorter guy-wires be fastened to the ground?
   d. What angle will the longer guy-wires make with the ground? The shorter guy-wires?

10. A shoreline observation post is located on a cliff such that the observer is 310 feet above sea level. When initially spotted, the angle of depression (the angle at which an observer needs to look down) from the observation post to an approaching ship was 4.8°.
   a. Draw a diagram of the situation and label each known and unknown quantity measure.
   b. How far out to sea was the ship when it was first spotted?
   c. After watching the ship for 47 seconds, the angle of depression was 5.9°. Assuming the ship was moving at a constant rate, estimate the ship's velocity.
   d. If the ship were to continue on its course at the same velocity, how much time from the second reading will elapse before the ship crashes into the shore?

11. An architect designs a building that has an overhang above a south-facing floor-to-ceiling window. This window is 10 feet high. The sunlight hits the ground at a 79.5° angle on the summer solstice (June 21) and 32.5° on the winter solstice (December 21).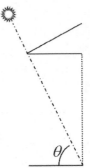
   a. Label the diagram with the known quantity measures.
   b. How long should the overhang project from the side of the building so that the sunlight hits just at the bottom of the window on the summer solstice?
   c. How long should the overhang project from the side of the building so that the sunlight hits just at the bottom of the window on the equinox (March 21 and October 21), when the sunlight's angle is 56° with respect to the ground?

12. An airport radio tower, which is 80 meters tall, is tracking a plane that is flying towards the airport, which is located at sea level. The plane is maintaining an altitude of 9.5 kilometers and flying at a speed of 482 kilometers per hour. Initially, the plane is 177 kilometers (the line of sight distance) from the radio tower.
   a. Draw a diagram of the situation and label each known and unknown quantity measure.
   b. What is the initial angle of elevation from the top of the radio tower to the plane? What is the initial horizontal distance of the plane to the radio tower?
   c. Define a function that relates the plane's angle of elevation from the top of the tower as a function of the time elapsed since beginning to track the plane.
   d. When is the plane directly above the tower?

13. While driving on a straight highway across the flat plains of eastern Wyoming, you notice a snow-capped mountain peak in the distance, directly ahead. When stopping at a gas station for coffee and snacks you use your electronic angle-measuring tool and find that the angle of elevation to the top of the mountain was 1.2637°. Later, after stopping at a rest area 25.0 miles further along the road, you find that the angle of elevation from there was 2.015°.
   a. Draw a diagram of the situation and label each known and unknown quantity measure.
   b. From the information above, determine the height of the mountain in feet.
   c. How far are you from the *top* of the mountain at each location (the line of sight distance)?
   d. You look to the south and notice a familiar mountain. On the map you find that you are 36 miles from the base of this mountain. You also know the mountain is 0.34 miles tall. What angle should the measuring tool return?

III. TRIGONOMETRIC IDENTITIES

14. a. Explain in your own words what the identity $\sin(2x) = 2\sin(x)\cos(x)$ conveys. (*Say more than "they are equal" - describe what it means to say that they are equal.*)

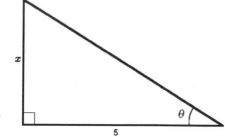

   b. Given that $\sin(x) = \frac{2}{3}$ and $\cos(x) > 0$, determine the values of $\cos(x)$, $\tan(x)$, $\csc(x)$, $\sec(x)$, and $\cot(x)$ without using inverse trigonometric functions to determine the value of x.

   c. Simplify the expression $\frac{\sin(\theta)}{1+\cos(\theta)} + \cot(\theta)$ until it is expressed as a single trigonometric function.

   d. Using the given right triangle, determine the values in terms of x of the six trigonometric functions with an input of θ radians.

   e. Use the sum and difference identities to prove $\cos(x - \frac{\pi}{2}) = \sin(x)$ and $\sin(x + \frac{\pi}{2}) = \cos(x)$ for all x.

15. For each of the following, complete the identity by simplifying. In other words, rewrite each of the following in its most simplified form.

   a. $\dfrac{\tan(x) \cdot \cot(x)}{\csc(x)}$

   b. $\sec(x) - \sec(x) \cdot \sin^2(x)$

16. Using the given geometric proof, fill in the missing lengths (A and B) and explain how each length is determined.

   $$\sin(x+y) = \sin(x)\cos(y) + \cos(x)\sin(y)$$

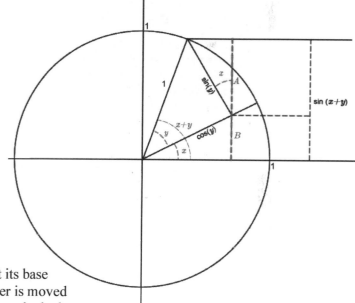

17. A 14 ft. ladder is leaning against a wall, such that its base is 12.9 ft from the wall. Then the base of the ladder is moved toward the wall into a new position, such that the angle the base makes with the floor has doubled from the angle of its initial position.

   a. Draw a diagram that represents the situation.

   b. Use the identity $\cos(2x) = 2\cos^2(x) - 1$ to determine the distance of the base of the ladder from the wall in its new position.

   c. Find the distance of the base of the ladder from the wall in its new position using another method.

18. Use the sum and difference identities and the Pythagorean identities to algebraically prove the following double angle identities.

   a. $\sin(2x) = 2\sin(x)\cos(x)$

   b. $\cos(2x) = 2\cos^2(x) - 1$

   c. $\cos(2x) = \cos^2(x) - \sin^2(x)$

   d. $\cos(2x) = 1 - 2\sin^2(x)$

19. Use the sum and difference identities to algebraically prove the following sum to product identities.
    a. $\frac{1}{2}(\sin(x-y)+\sin(x+y))=\sin(x)\cos(y)$ b. $\frac{1}{2}(\cos(x-y)+\cos(x+y))=\cos(x)\cos(y)$
    c. $\frac{1}{2}(\cos(x-y)-\cos(x+y))=\sin(x)\sin(y)$

20. Algebraically prove the following identities.
    a. $\tan(\theta)+\frac{\cos(\theta)}{1+\sin(\theta)}=\sec(\theta)$ b. $\tan^2(\theta)=\sec^2(\theta)-1$
    c. $\frac{1-\cos(2\theta)}{2}=\sin^2(\theta)$ d. $\frac{1+\cos(2\theta)}{2}=\cos^2(\theta)$

Pythagorean Identities:

$$\sin^2(\theta)+\cos^2(\theta)=1$$
$$\tan^2(\theta)+1=\sec^2(\theta)$$
$$1+\cot^2(\theta)=\csc^2(\theta)$$

Sum/Difference Identities:

$$\sin(\theta_1+\theta_2)=\sin(\theta_1)\cos(\theta_2)+\cos(\theta_1)\sin(\theta_2)$$
$$\sin(\theta_1-\theta_2)=\sin(\theta_1)\cos(\theta_2)-\cos(\theta_1)\sin(\theta_2)$$
$$\cos(\theta_1+\theta_2)=\cos(\theta_1)\cos(\theta_2)-\sin(\theta_1)\sin(\theta_2)$$
$$\cos(\theta_1-\theta_2)=\cos(\theta_1)\cos(\theta_2)+\sin(\theta_1)\sin(\theta_2)$$

Double Angle Identities:

$$\sin(2\theta)=2\sin(\theta)\cos(\theta)$$
$$\cos(2\theta)=\cos^2(\theta)-\sin^2(\theta)$$
$$\cos(2\theta)=2\cos^2(\theta)-1$$
$$\cos(2\theta)=1-2\sin^2(\theta)$$

Product to Sum Identities:

$$\sin(\theta_1)\cos(\theta_2)=\frac{1}{2}(\sin(\theta_1-\theta_2)+\sin(\theta_1+\theta_2))$$
$$\cos(\theta_1)\cos(\theta_2)=\frac{1}{2}(\cos(\theta_1-\theta_2)+\cos(\theta_1+\theta_2))$$
$$\sin(\theta_1)\sin(\theta_2)=\frac{1}{2}(\cos(\theta_1-\theta_2)-\cos(\theta_1+\theta_2))$$

Half Angle Identities:

$$\sin^2\left(\frac{\theta}{2}\right)=\frac{1-\cos(\theta)}{2}$$
$$\cos^2\left(\frac{\theta}{2}\right)=\frac{1+\cos(\theta)}{2}$$

IV. APPLYING TRIGONOMETRY TO NON-RIGHT TRIANGLES

21. The diagram shows some measurements taken for a new bridge. How long is the bridge?

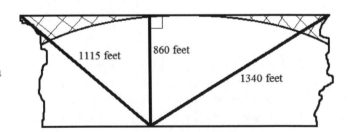

22. Three surveyors have placed themselves at three locations around the edge of a canyon and measure the angles between them. Surveyor A and Surveyor B are 1,359.2 feet apart. What is the distance between Surveyor A and Surveyor C? Surveyor B and Surveyor C?

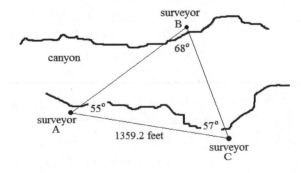

---

23. Standing on top of a building you use a laser measuring tape to measure your distance from the top and bottom of a nearby statue. How tall is the statue?

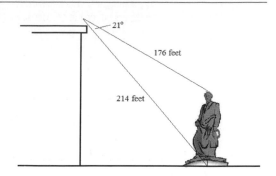

24. Find the value of *x* in each of the following triangles.

a.

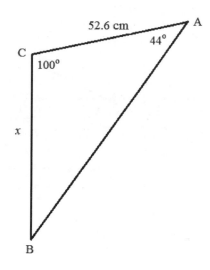

b.

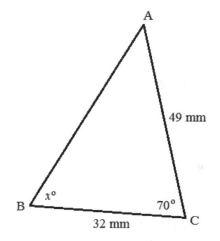

c.

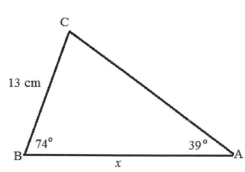

d.

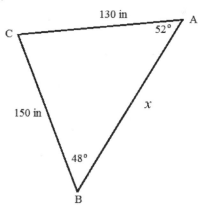

e.

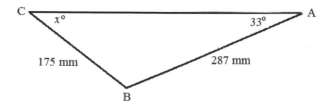

f.

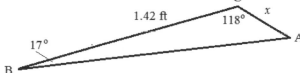

25. Find the area of each triangle.

a.

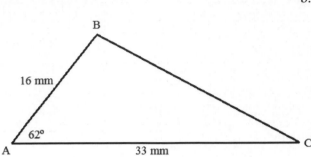

b.

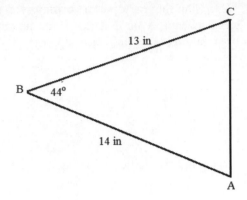

V. **THE LAW OF SINES**

The Law of Sines

$$\frac{a}{\sin(m\angle A)} = \frac{b}{\sin(m\angle B)} = \frac{c}{\sin(m\angle C)}$$

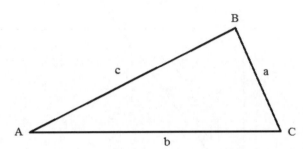

26. Use the Law of Sines to find the value of x in each of the following triangles.

a.

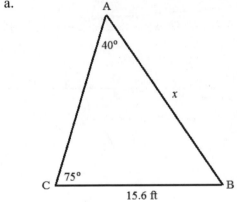

b.

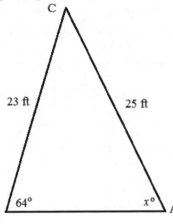

c.

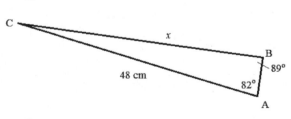

d.

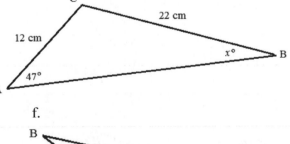

e.

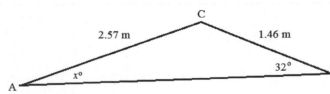

f.

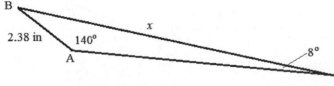

27. Find the length of the missing sides and the missing angle measures in the following triangles.
    a.

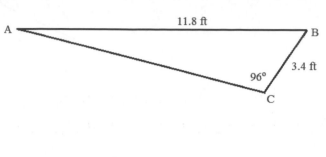

    b.

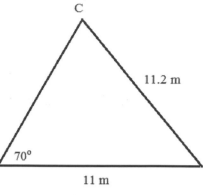

    c.

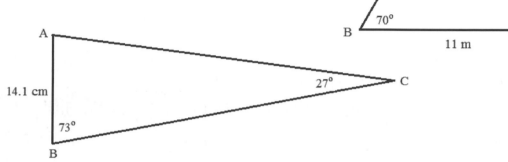

28. There are two possible triangles with the given measurements. Find the side lengths and angle measures of both possible triangles and draw each triangle approximately to scale.

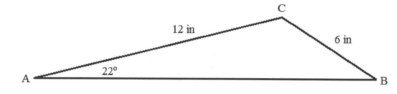

29. There are two possible triangles with the given measurements. Find the side lengths and angle measures of both possible triangles and draw each triangle approximately to scale.

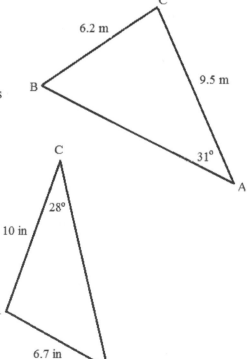

30. There are two possible triangles with the given measurements. Find the side lengths and angle measures of both possible triangles and draw each triangle approximately to scale.

VI. THE LAW OF COSINES

The Law of Cosines

$$a^2 = b^2 + c^2 - 2bc \cdot \cos(m\angle A)$$
$$b^2 = a^2 + c^2 - 2ac \cdot \cos(m\angle B)$$
$$c^2 = a^2 + b^2 - 2ab \cdot \cos(m\angle C)$$

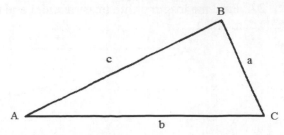

31. Use the Law of Cosines to find the value of x in each of the following diagrams.

a.

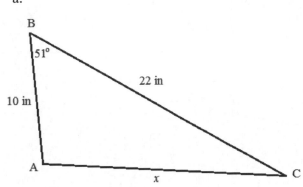

b.

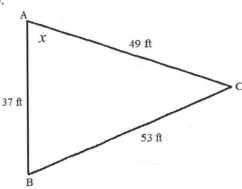

c.

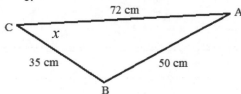

d.

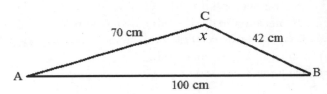

e.

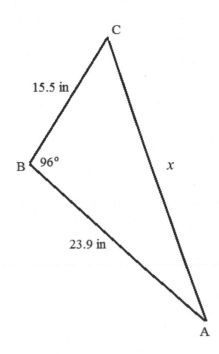

f.

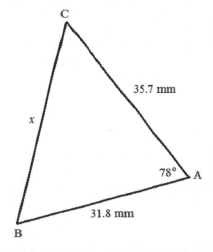

32. Find the missing side lengths and the missing angle measures.

a.

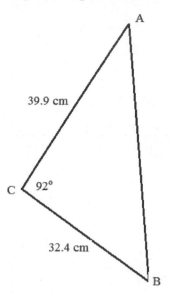

b.

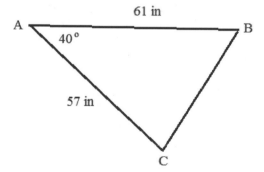

c.

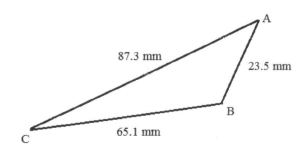

33. Find the value of x in each of the following diagrams.

a.

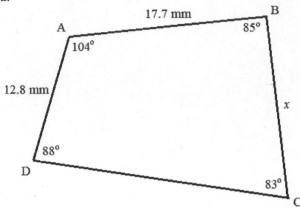

b.

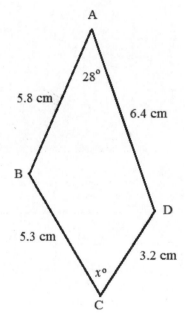

c.

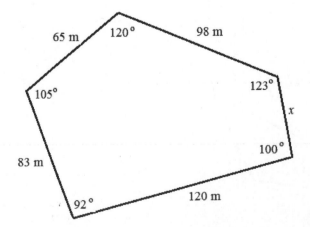